The Microscope and how to use it

BY

DR. GEORG STEHLI

TRANSLATED BY
WILLIAM A. VORDERWINKLER

DOVER PUBLICATIONS, INC., NEW YORK

This Dover edition, first published in 1970, is an unabridged and unaltered republication of the 1960 English translation of the work originally published in 1955 by the Franckh'sche Verlagshandlung, W. Keller & Co., Stuttgart, with the title *Mikroskopie für Jedermann.*

The present edition is published by special arrangement with Printed Arts Co., Inc., 419 Park Avenue South, New York, N.Y. 10016.

Standard Book Number: 486-22575-5

Library of Congress Catalog Card Number: 70-107464

Manufactured in the United States of America

Dover Publications, Inc.

180 Varick Street

New York, N.Y. 10014

CONTENTS

Foreword

The world can be thankful to the microscope for a great many things. How one-sided and limited our knowledge of nature would be if the trail-blazing scientist had not been able to discover new worlds by microscopic examination in the last two and a half centuries! There have been many discoveries of great significance in the battle for life against bacteria, the achievements of Pasteur, Koch, Behring and Ehrlich, to mention only a few.

In working with a microscope we learn to see nature with a second pair of eyes, and become better able to understand the sense and relationships of things in this great world. As students of nature, we find a completely new world unfolding itself. Observation of the life processes of seemingly simple plant life, unexpected views of the bodily construction of animals, of the swarming world in a drop of water—such insights into the secrets of nature are unforgettable and are part of the most precious, irreplaceable body of knowledge that modern man possesses.

Microscopy is also at least as intriguing as any other hobby of our time. Microscopy as a hobby does not require any special knowledge or skill. The cost, as this guide will show, is small—at least no more than any other hobby. This book is intended for readers who have had no previous training in microscopy, and who want to learn, by self-instruction, how to use a microscope.

This book is intended to serve as a methodical introduction into the active realm of microscopy. It introduces you to the construction and use of the microscope and the auxiliary tools, beginning with the simplest ones and acquainting you gradually and thoroughly with all the technical skills, so that you may continue to find new areas to use your knowledge.

You will experience from the very beginning and without laborious study the endless pleasures of looking through a microscope. With progressive practice you will soon learn the satisfaction gained from thorough, scientific work. If new difficulties arise, there are many specialized works available which will take you beyond this introductory knowledge, but which cannot be understood without this knowledge. There are also some publications which keep you posted on the latest advances in the field of microscopy and give all sorts of assistance by telling you about new works, new advances in chemicals and other materials as they become available. This is a very important point for a beginner, because in this manner he can get inexpensive study material which could otherwise be obtained only with difficulty and at considerable cost.

This book is perhaps more than introductory, but it leaves out such complicated subjects as dark-ground illumination, advanced microphotography, microprojections, etc.

The microscopy hobbyist, who is usually also otherwise occupied, can only seldom indulge in long, time-consuming procedures. For this reason particular attention has been given to modern short-cuts.

GEORG STEHLI

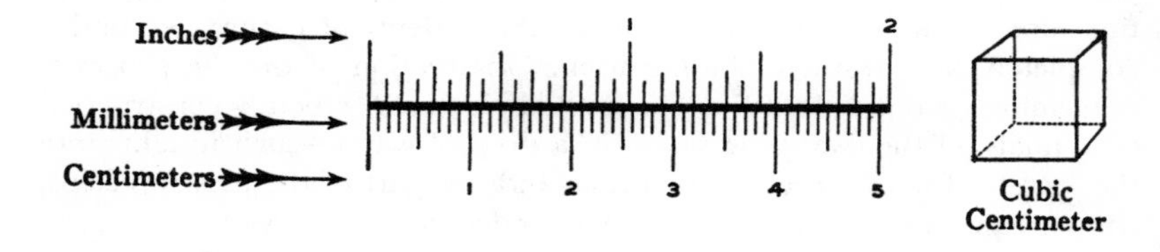

1. The Microscope and Essential Tools

A good microscope is the first and most important piece of equipment for your studies. Of course even the simplest and cheapest "magic tube" will unfold many secrets and new wonders; but everyone, once he has got beyond the simplest beginnings, will feel the need to delve further into the apparently puzzling things he sees, and to make more thorough observations requiring greater magnifications—in short, to see more. The disappointment is great when the microscope no longer "co-operates"—when the limits of its powers have been reached.

A really useful microscope is so made that it can be improved with the purchase of additional optical equipment. A good standard model is not very much more expensive than the widely used, but usually disappointingly small, student's microscope. It offers the great advantage that by and by, as your purse permits, it can be made into a complete scientific instrument. A simple and therefore inexpensive instrument is sufficient to begin with, but it should be a model that can be elaborated. For ordinary purposes a $\frac{1}{2}$ in. and a $\frac{1}{6}$ in. objective and a $\times 5$ eyepiece will suffice, but it is advisable to purchase a condenser at the outset. An instrument which is not too expensive and may later be built up is shown in Fig. 1.

HOW A MICROSCOPE IS CONSTRUCTED

The microscope in Fig. 1 rests on a horseshoe-shaped base or foot from which a low, solid column rises. To this column the upper part of the microscope is joined with a simple, quite tight hinge, which enables the body tube of the microscope to be tilted to an angle of 45°; this permits a comfortable position for the head of the observer. A short distance above the hinge is a roomy, square object stage, on which the slide is firmly held by two spring clips. Above the stage there is a sturdy limb which rises above the center of the stage; this holds the tube that carries the lenses. The objective is screwed to the lower end of the tube, and the upper end holds the eyepiece.

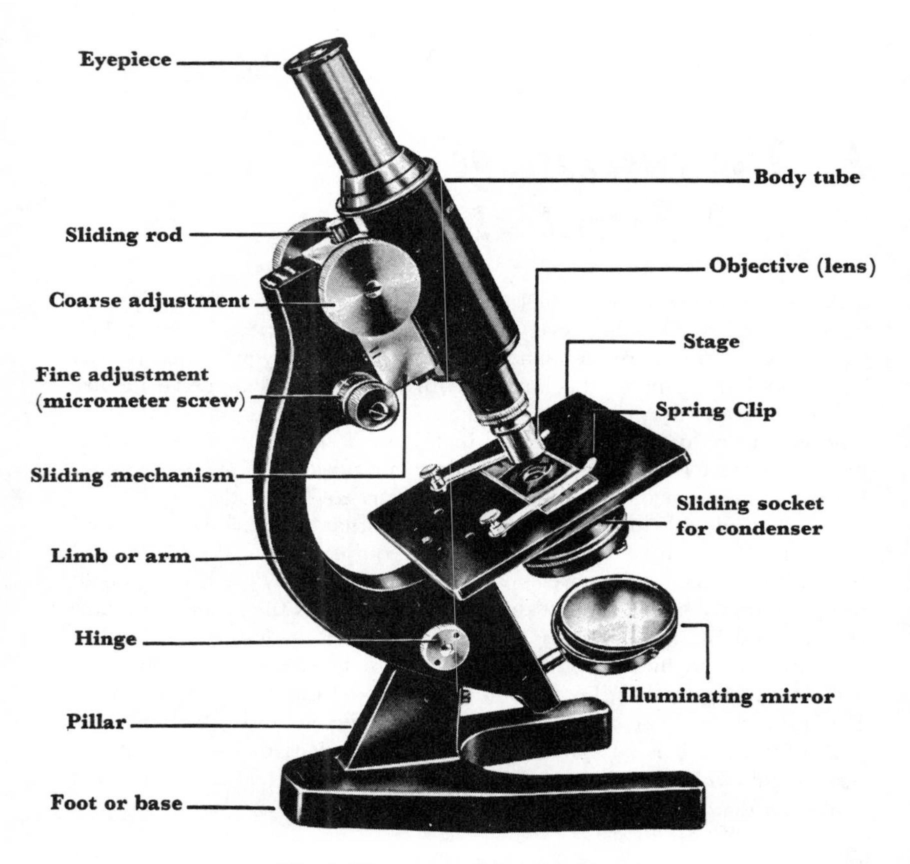

Fig. 1. The parts of the microscope.

For coarse focusing of the microscope there is a rack and pinion adjustment controlled by two large knurled knobs: this moves the tube up and down rapidly. The fine adjustment is made by a second motion controlled by two smaller knobs, the micrometer screws.

With the fine adjustment it is possible to move the tube a minute distance, measured in hundredths of a millimeter. The mechanism for this adjustment is contained within the top of the limb (arm).

The optical arrangement of the microscope is shown schematically in Fig. 2. You see here a cutaway view through the optical axis of an

Reichert (Viennese) "Biozet" microscope with micro-drawing attachment

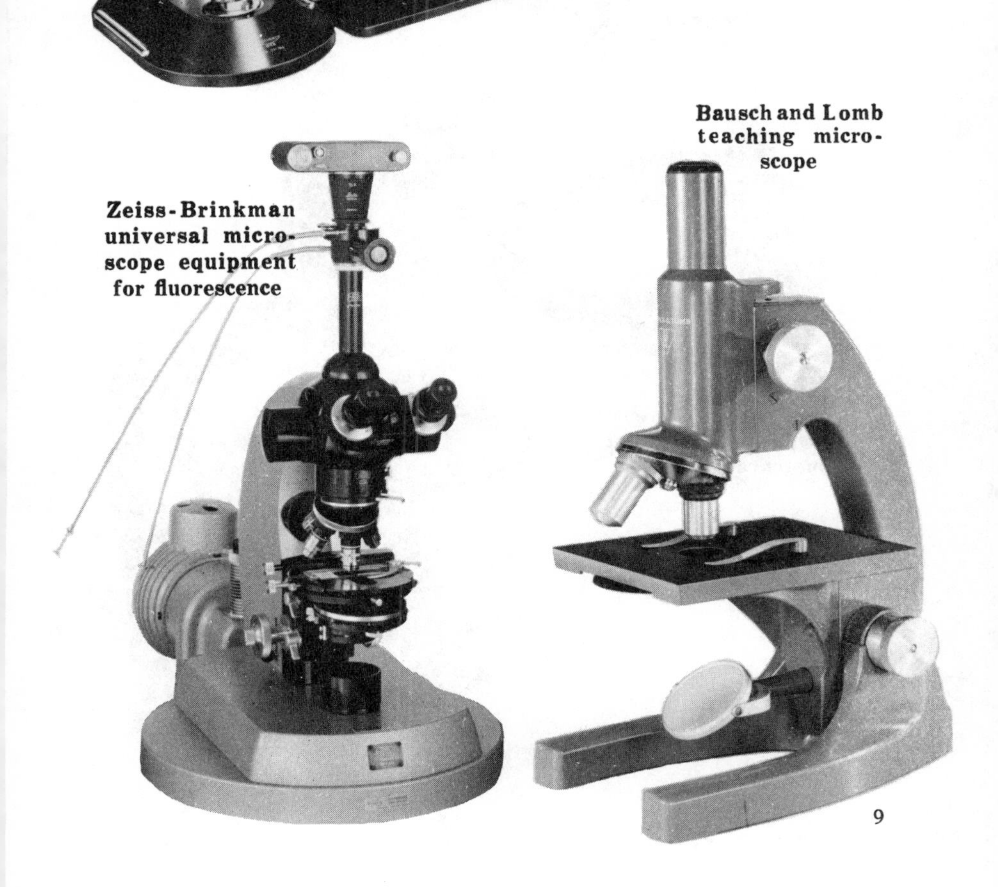

Bausch and Lomb teaching microscope

Zeiss-Brinkman universal microscope equipment for fluorescence

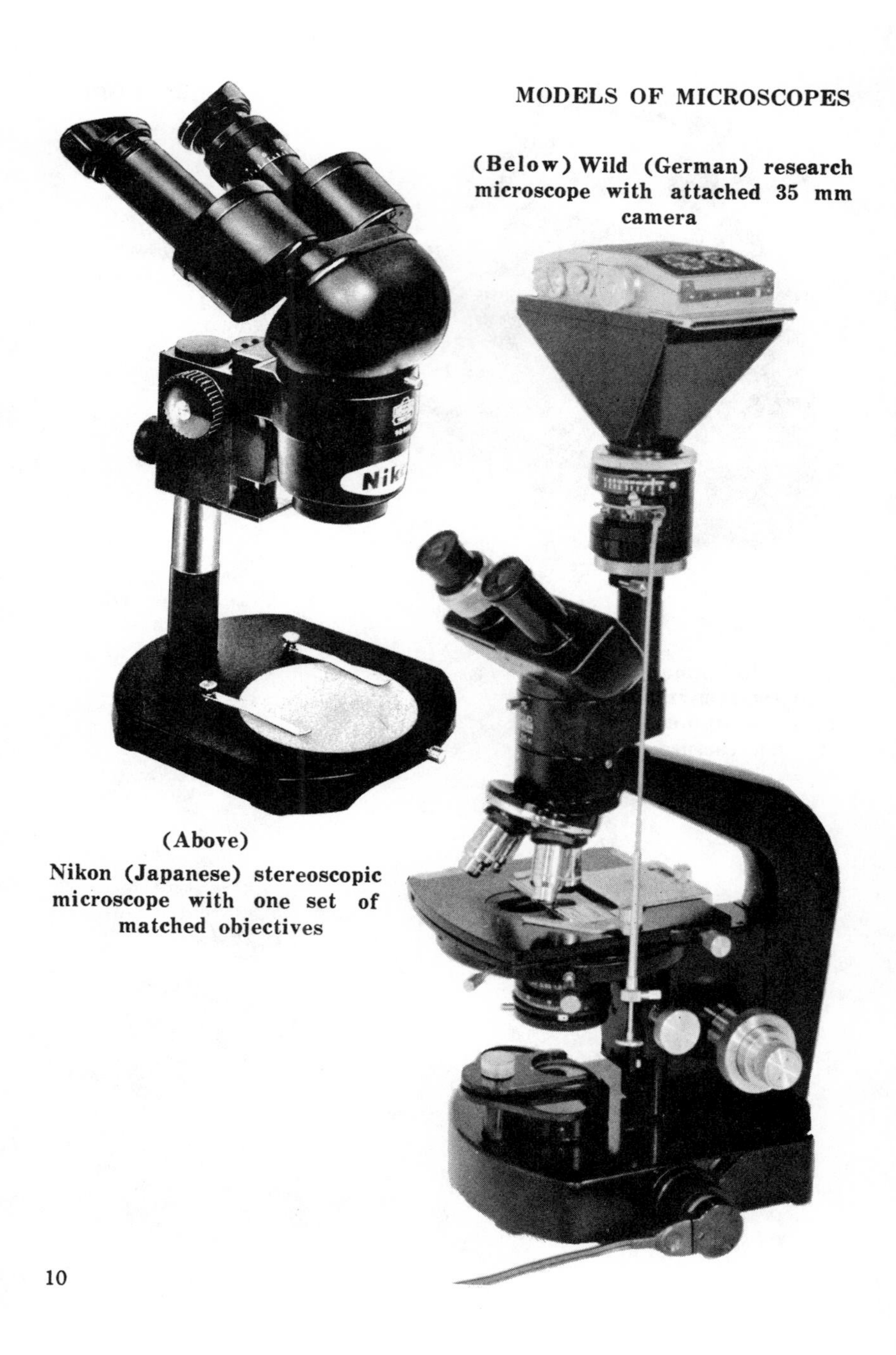

(Below) Wild (German) research microscope with attached 35 mm camera

(Above)
Nikon (Japanese) stereoscopic microscope with one set of matched objectives

MODELS OF MICROSCOPES

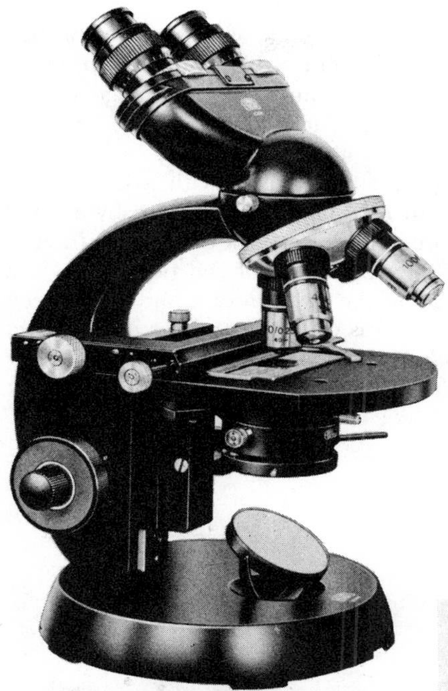

(Left) Carl Zeiss laboratory microscope with inclined binocular tube

(Left, below) American Optical laboratory student microscope

(Below) Swift teaching microscope

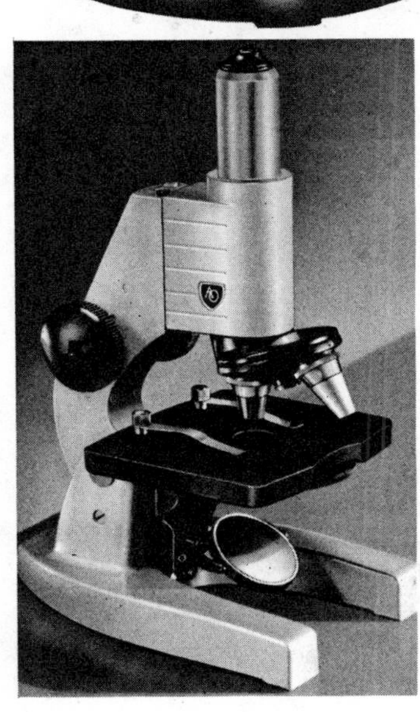

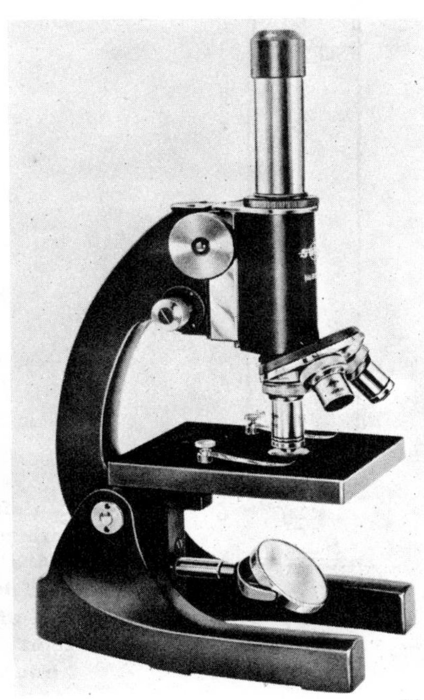

assembled microscope, represented here by line *x* which passes through the central point of the lenses. *C* is the system of lenses in the objective. These lenses are compound or achromatic, and give an image that is without colored edges. Each lens does, in fact, consist of two or three separate lenses made of different kinds of glass.

Beneath the objective is the stage *t*, a metal plate with a hole in the center, on which the slide preparation *P* is placed and lighted by a small beam. The objective *C* is screwed into the lower end of a metal tube *T* which is blackened internally. At the upper end of the tube there is a second system of lenses *A*. It consists of the two lenses *a* and *d*, between which there is usually a diaphragm, and is called the ocular or eyepiece, because it is here that the eye (*oculus*) of the observer is applied. Under the stage there is an illuminating mirror *S*, which reflects the necessary

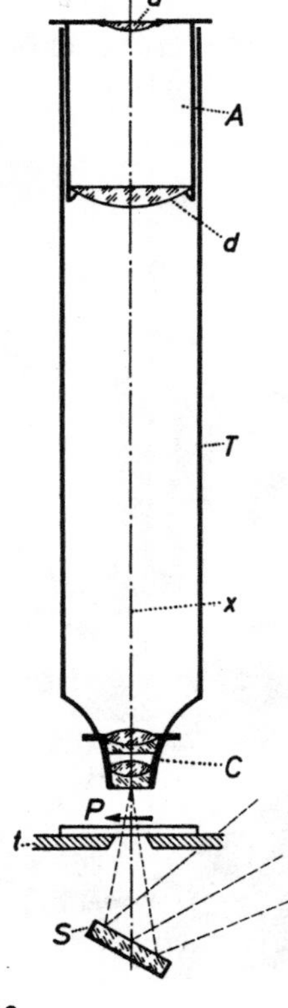

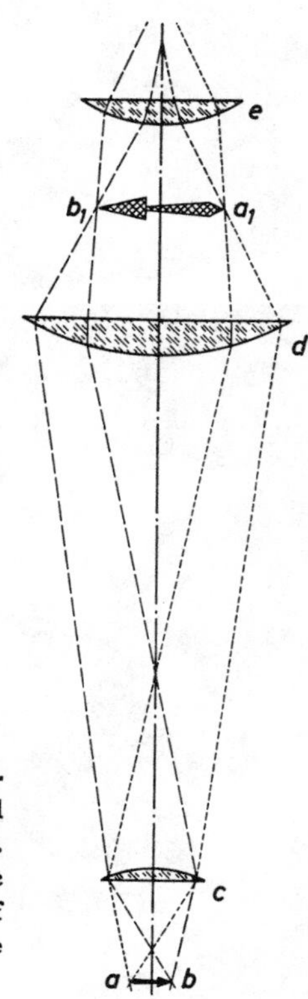

Fig. 2. (left) Cross-section through the optical part of a microscope. Fig. 3. (right) Schematic diagram of the path of the light rays in the microscope.

light through the hole in the stage on to the slide. It follows that the object being examined must be transparent. As Fig. 2 shows, the optical axis x (the tube axis) passes through the exact center of all parts of the instrument. This is absolutely necessary if the microscope is to deliver sharp, undistorted images.

How the optical components work is shown in Fig. 3. The objective c is represented by only one lens for the sake of simplicity. It projects an exact and therefore photographable image of an object, which is represented by lines a to b. The ocular lens d breaks up these rays which in the diaphragm behind d are united in an enlarged but reversed image b_1 to a_1. This magnified image is seen through lens e of the eyepiece as if it were seen through an ordinary magnifying glass; the observer sees a repeated, if only weakly magnified, clear image of the object. Because the common Huygenian eyepiece cannot "erect" the image, the view seen in the microscope is reversed. It follows that if you wish to move the examined object to the right or left, or up or down, you must move the slide in the opposite direction to that desired. This may take a little getting used to.

Like any good microscope, the Humboldt may be equipped with several objective lenses (e.g. 1-in., $\frac{1}{2}$-in. and $\frac{1}{6}$-in.) and several eyepieces (e.g. $\times 5$ and $\times 12$). The objective lenses are corrected to suit an exact tube length; in the case of the Humboldt this is 170 mm.[1] Objectives and eyepieces can be combined in many ways, so as to give a great range of magnifications (explained in a chart which comes with the instrument). The total magnification of a microscope is determined by a combination of the objective and eyepiece, and is the product of the two magnifications. An objective magnifying 30 diameters, used with an eyepiece of 6 diameters, would give a total magnification of 180 diameters. The magnifying power of the objectives and eyepieces depends upon their focal lengths, and a system magnifies more highly as its focal length decreases.[2]

At this point it might be well to mention that, contrary to common opinion, strength of magnification does not determine the value of a microscope. Much more important than the power of magnification is a microscope's so-called "resolution". This is the power of an optical system to separate minute dots or lines so that details can be distinguished. With a good microscope and the most advantageous lighting the best

---

1 This so-called mechanical tube length may not be changed, not even if a nosepiece is afterwards screwed to the bottom of the tube (see Fig. 4). Here you can help by taking out an intermediate ring which corrects the length by exactly the height of the nosepiece.

2 In the Humboldt microscope the objective, tube and nosepiece all have international standard threads. This makes it possible to use other makes of objective and eyepiece and, by the same token, the Humboldt lenses may be used on other microscope stands.

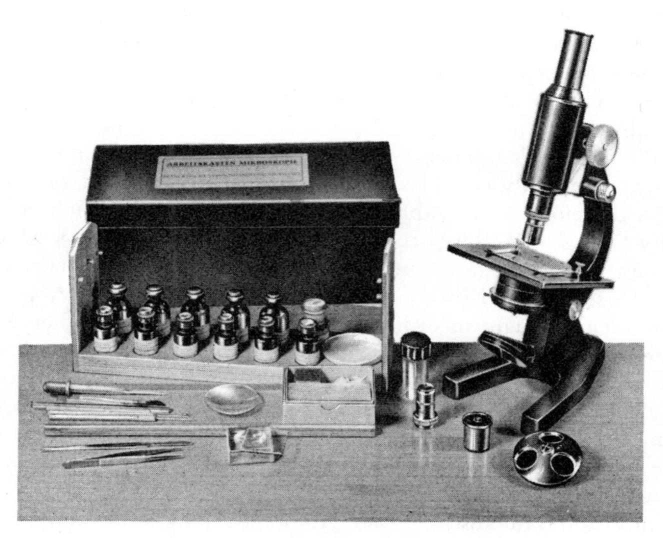

Fig. 4. Microscopy work kit.

attainable detail lies around 0.2μ ($1\mu = 1$ micron $= 1/1000$ mm.). The tiniest bacteria, which are a little bigger than 0.3μ, can be seen with the help of an exceptional objective—the so-called "oil-immersion lens". Viruses, on the other hand, which are considerably smaller, cannot be seen with an optical microscope at all.

The powers of detail of our eyes are much less than that of a usable microscope—and that is why we can see more with the microscope than with the naked eye. The objective lenses are primarily responsible for the detail rendered, whereas the ocular lenses do not affect detail to any great extent. The objective lenses are therefore the most important (and sensitive) parts of the microscope and must be handled with special care.

Many beginners make the mistake of trying to get immense magnifications by the use of particularly strong eyepieces. It is possible of course to combine a 1/12-in. objective lens with a $\times 25$ eyepiece, for example, and in this way arrive at a magnification of 2,500 diameters. The detail, however, is not increased by the powerful eyepiece, and you cannot see more than with an eyepiece of about $\times 12$. In addition, there is another factor. With unnecessarily powerful eyepieces the image becomes very weakly lighted, the structures no longer appear clear and sharply defined, and there is danger of optical illusions.

OBJECTIVE	EYEPIECE								
	Simple Magnification			× 5	× 6	× 8	× 10	× 12	× 14
	Description			50	40	30	25	20	17
	Focal Length (mm.)			5	6	8	10	12	14
Description	Focal length (mm.)	Aper-ture	Simple magnifi-cation	× 5	× 6	× 8	× 10	× 12	× 14
1½ in.	36	0.10	× 5	25	30	40	50	60	70
1 in.	25	0.20	× 7	35	42	56	70	84	98
⅔ in.	16	0.30	× 10	50	60	80	100	120	140
½ in.	10	0.45	× 18	90	108	144	180	216	252
¼ in.	6	0.60	× 30	150	180	240	300	360	480
⅙ in.	4	0.82	× 50	250	300	400	500	600	700
⅛ in.	3	0.85	× 60	300	360	480	600	720	840
× 100 (oil-immersion 1/12 in.)	1.8	1.30	× 100	500	600	800	1000	1200	1400
× 115 (oil-immersion 1/16 in.)	1.5	1.30	× 115	575	690	920	1150	1380	1610

The remedy is to find out with which eyepiece you can work effectively with a given objective. The aperture of every objective is designated by a number—either on the objective itself or on a list provided by the manufacturer. The general rule is that the total magnification should never be greater than 1,000 times the number of the aperture. For a 1/12-in. objective (the so-called "oil immersion") with an aperture of 1.30 the highest eyepiece which you can use is ×13. A more powerful eyepiece would lead to "empty magnifications", which can be used only for special purposes, such as the measurement of very small bodies. But there is also a limit at the other end of the scale. The total magnification should not go below 500 times the aperture of the objective, or the detail potentialities would not be fully put to use. This range of magnification between 500 and 1,000 times the aperture is referred to as "usable magnification".

Besides the microscope you will require a little arsenal of preparing instruments, glass equipment and other aids. It is difficult to give the beginner general advice, because the tools necessary for the different branches of microscopy are diversified. Therefore, we will suggest here equipment that will be sufficient only for the very beginning. It is a good idea to keep this equipment all together in a box or "kit":

Standard microscope slides or slips (1 in. × 3 in.). These can be purchased cheaply at any shop where microscopes are sold. The blank slides are traditionally called *slips*, and those with mounted objects *slides*, but these terms are becoming interchangeable.

Cover-glasses or coverslips are extremely thin, circular or square pieces of glass for covering objects to be examined. No. 2 or 3, ¾ in. in diameter, are suitable for general purposes, but very thin coverslips are unnecessarily fragile for ordinary work.

Watch-glasses and glass blocks (so-called "salt-shakers") for holding small masses being prepared. You should have 5 watch-glasses about 2 in. in diameter.

Larger stocks of materials, such as bits of twigs, leaves, organs of animals, etc., which are stored in inexpensive preparation-glasses. Old medicinal jars, small preserve jars, etc., will serve the purpose. (Of course you can never have too many glass containers, as you will soon find.)

A thin glass rod about 8 in. long with hemispheral ends, for transferring drops of liquids to slides.

A pair of tweezers made of good steel, for grasping cover-glasses and other objects which would be awkward to handle with the fingers.

A needle-holder with several sized preparation or "setting" needles to tear apart the materials for examination. Ordinary sewing needles may be used in a needle-holder, or permanently fixed into any convenient handle such as a wooden pen-holder. In this way you can make a set of good preparation needles yourself.

Two fine camel's-hair brushes, which are very handy for transferring objects to slides.

A pipette, a glass tube of small diameter, with which small quantities of fluids can be drawn up in order to isolate them for examination. Such pipettes can easily be made. Take a larger glass tube and heat it in the middle in the upper part of a spirit or gas flame, turning it constantly until it becomes red all over. Then pull both ends of the tube apart. Break the thinned-out ends or file them apart. In this way you will get two pipettes, which you can cap with little pieces of rubber for easier handling. This eliminates the necessity of holding your thumb over the

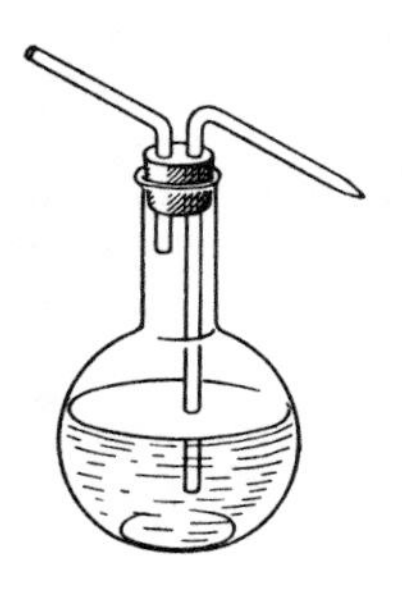

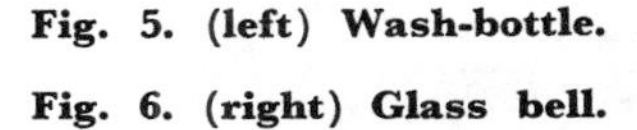

Fig. 5. (left) Wash-bottle.

Fig. 6. (right) Glass bell.

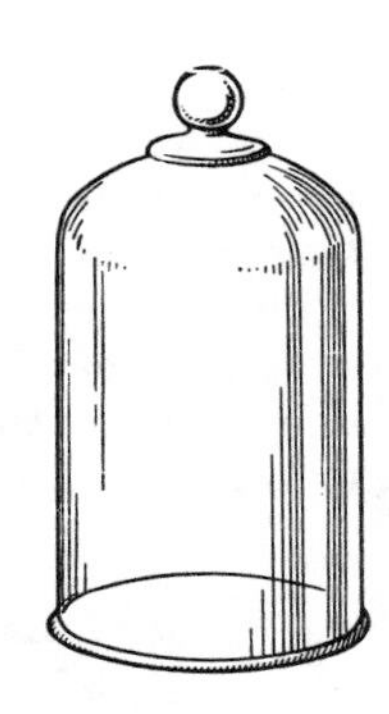

end when it is full. An eye-dropper or fountain-pen filler will serve if you are unable to make your own pipette.

To begin with you should get yourself a little wash-bottle (Fig. 5), a spirit lamp with a tripod and wire-netting (Figs. 7 and 8) and some glass bells (Fig. 6), for which purpose preserve jars may also be used. You will need the glass bells to protect open preparations from dust. You can make most of this apparatus yourself. A spirit lamp may be made easily from an ink bottle into which a metal sleeve is fitted to serve as a wick-holder. The tripod can be constructed out of heavy iron wire. The wash-bottle is a little more work. A small flask, perhaps a medicine bottle with a wide mouth, can be converted into a wash-bottle. Take a cork into which you have drilled two holes and insert in the holes two glass tubes, bent over a flame, like those in Fig. 5. If you blow into one tube, the other (which has to be drawn to a fine point) squirts out a thin stream of water. If a greater quantity of water is required, pour it out of the other tube. Wine-glasses, with the stems broken or otherwise, make very good glass bells.

Other tools you will need can be kept in a cigar box padded with cotton-wool. They include:

A scalpel, a pointed knife of good steel, set firmly in a handle. This will serve to cut smaller objects into shape or to dissect larger objects.

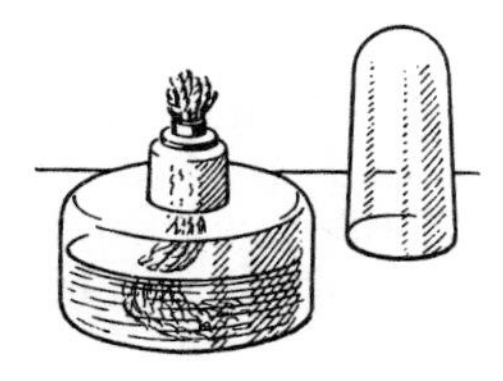

Fig. 7. (left) Spirit or alcohol lamp. Fig. 8. (right) Tripod made of heavy wire.

A lancet, a needle whose point has the shape of a tiny lance. This is the preferred cutting instrument of the zoologist, because both edges of the point are sharpened. The botanist uses it for transferring larger sections from liquids to slides.

Besides these tools every microscopist needs a sharp pocketknife, a strong pair of scissors and a small pair of embroidery scissors. It is advisable to buy a complete microscopy kit at the beginning, with all the preparation tools in their proper places (Fig. 9).

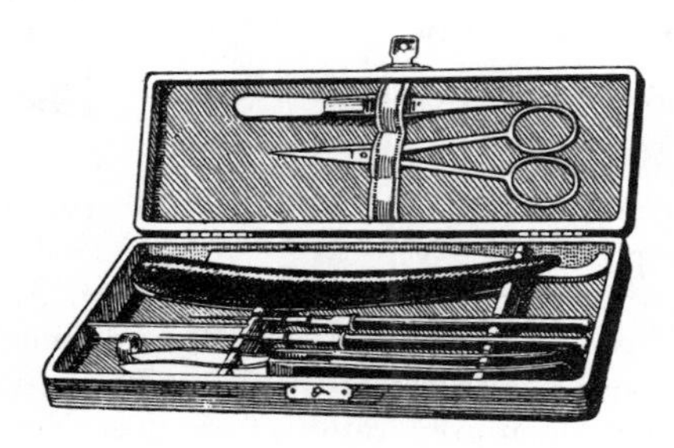

Fig. 9. Microscopy tools.

For chemicals, the basics are: xylol, glycerin jelly and either an artificial resin such as Permount, or Canada balsam, for mounting; methylene benzoate; glycerin; and eau de Javelle (or "Milton") to bleach the preparation. Because the last-named, like tincture of iodine, bites into metal, great care is advised (with the microscope!).

Dye preparations in temporarily usable solution: tincture of carmine, Grenacher's borax, Delafield's hematoxylin, eosin, safranin and Mayer's acid hemalum. Also Löffler's methylene blue.

Outside of these chemicals you will need alcohol. You can do without absolute alcohol, except for fixative purposes. For ordinary examinations, industrial spirit is sufficient, if it is 95%. For closer examinations, to add to dye preparations, to preserve plants and animals, and for dyhydration of objects for permanent preparations, you will use isopropyl alcohol, which is obtainable as 100%, completely free of water. Because it cannot be drunk, it is very much cheaper than absolute alcohol.

You can prepare alcohol of a lower concentration by mixing 95% alcohol with distilled water. Follow these instructions (if a small error is made, it makes no difference):

Fill into a graduated cylinder as much 95% alcohol as your solution should amount to. (For example, take 50 cubic centimeters [cc.] to get 50% alcohol.) Then fill the graduated cylinder with distilled water to a height of 95 cc. If you have 70% alcohol and want 50%, take 50 cc. and then fill with distilled water to 70 cc.

For most purposes you should keep a supply of 35, 70, 80, 90 and 95% alcohol (for the so-called alcohol stages of dehydration). For 100% alcohol, which you will also need, use water-free isopropyl alcohol. To store the alcohol stages it is best to use 4-oz. or 8-oz. bottles which can be sealed tightly. The liquid should not touch the corks. Alcohol must never be allowed to stand open for any time; it is combustible, evaporates very rapidly, and a high-test alcohol has the unpleasant quality of absorbing water from the air, thereby causing an undesirable dilution.

For dye preparations, you will need acetic acid to be used in an alcohol solution.

Hydrochloric acid, available in 37% solution, will also be needed.

All reagents must be properly labeled. In order not to tip them over, place them in an uncovered cigar-box.

In addition, your work table should have two containers of fresh water for cleaning the preparation tools—a dirty glass rod or a dirty needle could introduce foreign substances into the preparation and spoil everything.

There should be a box of labels to identify permanent preparations.

Elderberry pith and cork should be on hand as well, to hold small objects while cutting them.

A few sheets of filter paper complete the first equipment which the beginner requires for his tasks.

Everyone who works thoroughly keeps records. Your notes should begin when your materials are collected, and should continue through all stages of preparation. It is advisable to keep an individual page for each object, writing in remarks as to the origin of the materials, the manner of preservation and preparation, and also the number of the case in which the preparation is to be placed. Repeat the name of the object in your notebook index.

For these records you can use a bound book, a paper-covered notebook of good writing paper, or a loose-leaf notebook. Those who prefer loose pages may, instead, use index cards (which fit into a box) or cards which fit into a binder.

CLEANING OF COVERSLIPS, SLIDES AND LENSES

Coverslips and slides must be meticulously cleaned before using. Cleaning coverslips requires some special practice, or you will break them by the dozen! This work is made much easier if you keep a number of coverslips and slides in advance in a glass jar filled with spirit. If you need a cover-glass, get it out of the jar with a pair of tweezers, then grasp it with a fine, frequently-washed linen cloth, and rub it with a very

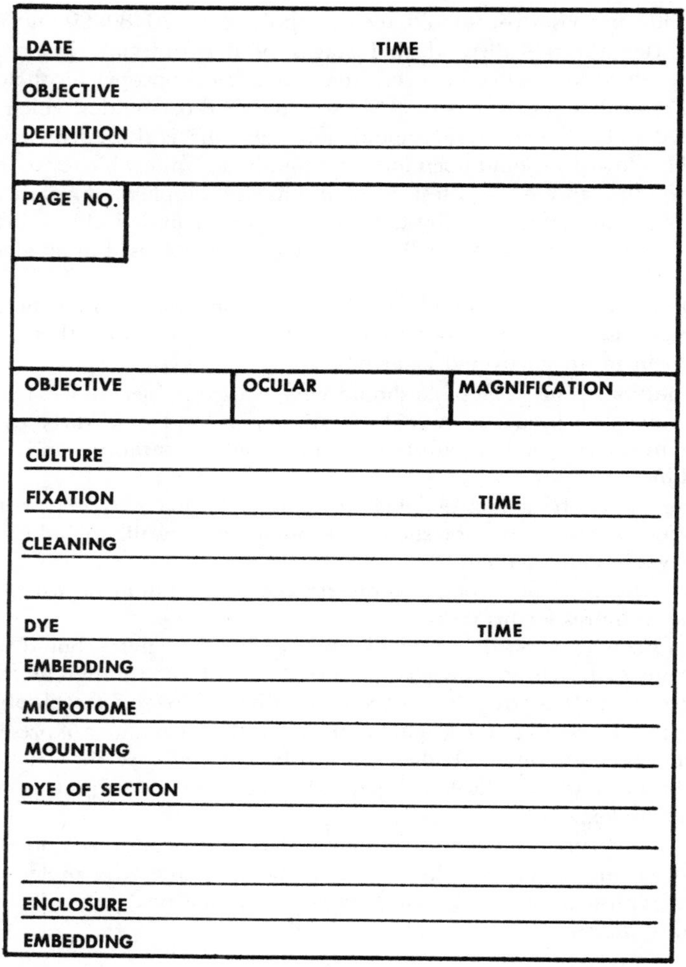

DATE TIME

OBJECTIVE

DEFINITION

PAGE NO.

OBJECTIVE	OCULAR	MAGNIFICATION

CULTURE

FIXATION TIME

CLEANING

DYE TIME

EMBEDDING

MICROTOME

MOUNTING

DYE OF SECTION

ENCLOSURE

EMBEDDING

Fig. 10.
Reproduction of a card for notations and drawings.

slight pressure. When the glass is thoroughly clean, grasp it by the edges between the thumb and forefinger of your other hand and place it on the preparation. Never touch the surface of the coverslip with your fingers, or greasy fingerprints will get on it, which will interfere with the examination of the object.

Clean your slides in the same manner; less caution is required with the heavier glass slide. The slides must not be handled either, except by the edges. Cleanliness is the alpha and omega of all microscopic work!

Objectives and eyepieces must be handled with special care. Here the strict rule is: "fingers off the lenses!" They must be protected from dust and must not be squeezed or scratched. For cleaning, first blow off all loose dust and then use clean, washed linen moistened with a drop of benzine. Remove lint left by the cloth with a clean camel's-hair brush.

The lens systems should not be unscrewed. In exceptional cases the upper lens of the eyepiece may be unscrewed for cleaning—the objective lenses, however, should be taken apart only by a trained optician or by the manufacturer.

When using chemicals be careful to avoid getting the slightest trace on the objective. If, in spite of this, the front lens of the objective should become smudged, wash it at once with a soft cloth moistened with water and glycerin jelly, or with a little benzine for removing mounting resin. Quite large cover-glasses are the best protection against such mishaps.

CARE OF OTHER TOOLS USED IN MICROSCOPY

Metal equipment (knives, scissors, needles, etc.) must never be left lying in a wet spot, but, especially after they are used for cutting, must be dried with a piece of soft linen and cleaned with a piece of chamois. If the work is to be interrupted for a time (weeks or months) metal tools should be coated with a thin coating of vaseline.

The microscope should be checked after each task to see that everything is in order.

2. How to Use the Microscope

The work table of the microscopist should be of the largest possible size. You will need the surface not only to set up the microscope, but also to prepare, stain, draw, and perform many jobs of that nature, all of which require a certain amount of equipment. All utensils necessary for the proposed investigations should have a place on the table. However, you must avoid using too much equipment: test tubes, staining vessels, slide cases, bottles, books, etc., which are not necessary for the work being done should be put on a shelf or in a cabinet, so that enough clear space remains.

The best seat is an adjustable stool.

The work table must be at a window, if you intend to work in daylight. Curtains should be removed from the window because they interfere with microscopy and shut off an unbelievable amount of light.

To set up the microscope, take it from its case, grasping it—as you must do every time you move it—by its limb. Now screw into its lower end the weakest objective. By carefully raising or lowering the tube the objective is adjusted in such a manner that the front lens (the outer, visible lens of the objective) is set about $\frac{1}{4}$ in. from the stage. You can check this by looking at the objective from the side, being careful that nothing comes in contact with it. The least little scratch will ruin the lens.

Before putting the eyepiece in place, look into the tube and try by turning the microscope mirror up and down, and from side to side, to light up the back of the objective so that it is evenly illuminated all over. Now insert a low-power eyepiece, through which you will get a round—and if the lighting is properly adjusted—evenly lighted "field of vision".

The microscope mirror consists of a flat mirror on one side, and a concave one on the other. The question: "when do I use the flat mirror, and when is it advisable to use the concave mirror?" is often asked and seldom answered correctly. Certainly it is difficult to give an answer which holds good in all cases. The concave mirror gives a brighter field of vision, but with the use of the flat mirror the structure of examined objects usually appears sharper. If the microscope is not provided with its own lighting device or with a condenser, then you must use the concave mirror with magnifications over 100 diameters, because with the flat mirror the image would be too weakly lighted. On the other hand, if there is a condenser you will work exclusively with the flat mirror.

The condenser is indispensable when using high and extremely high magnifications because only with the help of the condenser does the image become light enough for close examination. The condenser is fastened in a socket between the stage and the mirror and may be adjusted to a higher or lower position, as required for the objective being used (see Fig. 11).

Fig. 11. The two-lens condenser with diaphragm for the Humboldt microscope.

The condenser is equipped with an iris diaphragm, and in many cases also has a holder for a filter. When using very weak objectives the condenser is removed by pulling out its socket, or by swinging it to one side on a pivot, according to the construction of the microscope.

ADJUSTING THE PREPARED OBJECT

When you purchase your microscope you will probably find a sample preparation included, usually diatoms or butterfly scales. If such a preparation is not given, you can prepare a sample according to the instructions in Chapter 3.

Place the prepared slide on the stage, with the coverslip on top, in such a position that the area to be observed lies exactly above the round opening in the stage. Now look at the objective from the side and turn

Fig. 12. The condenser of Fig. 11 seen from the underside. The diaphragm, shown half open, is opened or shut with the upper knob. The frame which is swung out serves to hold a ground glass disc to cut down bright light and a blue disc to make artificial light resemble daylight. It can be swung out in one motion.

the coarse adjustment thumbscrew to lower the tube until the front lens of the objective is only a few millimeters above the coverslip. Only now may you look through the eyepiece and at the same time turn up the tube. If the lighting is correct, and the preparation is in its place you can look sharply at the visual range. At a certain height there will suddenly appear a blurred image of the prepared object. If the tube is raised again, the image disappears, and if the tube is carefully lowered again, the image appears once more. Therefore it will be seen that only with a certain distance between the objective lens and the prepared object does the image appear clear enough for examination. This distance becomes shorter as the strength of the objective increases. You must therefore work very carefully with the more powerful objectives, so that the costly front lens does not even lightly touch the slide. It is for this purpose that the microscope is provided with the fine adjustment.

It may sometimes happen to the beginner that no matter what pains he takes he cannot get an image into his field. What can be the cause of this? In the first place it is possible that the examined object is not lying exactly under the objective; you must accordingly move the object. Then again it may happen that when raising the tube, the object has "slid away" from the only spot where the image can appear; in such a case, you must again start your adjustments from the beginning.

Remember as the most important basic rule: raise the tube until the image appears. If you do the opposite, that is, lower the tube, then you let yourself in for a double danger: it is all too easy to touch the slide with the objective and destroy the preparation—and if the objective is not totally destroyed, it may become so scratched that it will no longer show a clear image.

The coarse adjustment ends as soon as the image appears in the visual range. For the fine adjustment, use the small micrometer knob. By turning it back and forth, try to get a sharply-defined image of the preparation. The micrometer adjustment raises or lowers the tube a very small distance with each turn. When using it there is not so great a necessity for care as with the coarse adjustment.

HOW TO GET THE BEST LIGHTING

With your later work you will often need the microscope for no more than brief, superficial examinations, which will serve as a first orientation, for example, in the control of dyeing procedures, for rapid sorting of a Plankton sample, or for the selection of a particularly thin section. In such cases you need not rack your brains greatly about illumination. With proper adjustment of the mirror you try to get the most evenly illuminated field: adjust the condenser farther away with weaker objectives; with more powerful objectives raise it almost to the top of its limits. The brightness is regulated with the adjustment of the condenser diaphragm, or if there is no condenser, by adjusting the stage diaphragm. As a light source for this sort of examination, daylight is sufficient.

You must pay much more attention to proper illumination when you are examining prepared objects carefully and need to note the smallest details. The magnifying powers of an objective are not necessarily the same as the actual magnification attained. Whether or not the full magnification is achieved depends greatly upon the lighting.

Daylight cannot even be considered for finer microscopic examinations. You need an artificial source of light even for a proper adjustment. In the simplest cases, take an ordinary electric lamp with a strong unfrosted incandescent bulb. A sewing-machine light will also do good service. Best, of course, is a complete low-voltage microscope light with a condenser and a light diaphragm, but this is quite expensive to purchase.

In any case it is important that the bulb is not frosted and that between the bulb and the microscope condenser a frosted glass disc or filter can be installed. The disc is placed in the filter frame or placed between the bulb and microscope mirror in a sort of holder. In an emergency a piece of tracing paper may be used instead of the glass disc.

First, focus the preparation with the flat mirror, being careful from the beginning to attain an even illumination throughout the field (see page 25). Now, take out the frosted glass and the slide. Look through the eyepiece and slide the condenser up and down until you see the image of your light source (the glowing filament of the bulb) clearly in the field; usually a quite high adjustment of the condenser is required. Then remove the eyepiece and partly close the iris diaphragm. Now if you look through the tube you will see on the back lens of the objective the image of the condenser diaphragm. Open the condenser diaphragm far enough so that its edge is covered with the rear objective lens. Then replace the eyepiece and the frosted glass filter, put back the slide, and you may begin your observations.

If all the different steps have been taken in an orderly manner, you will find a wonderfully clear and evenly bright field. Smaller corrections may still be made with a small adjustment of the mirror and very careful changes in the diaphragm opening. Thus, for viewing colored objects, especially dyed objects, you open the condenser diaphragm a little to make the colored image as bright as possible. When viewing uncolored objects the opposite is often advisable; closing the diaphragm brings out the details in full sharpness. Complicated as the lighting adjustments may seem when you read these instructions, after some practice you will succeed in finding the optimum lighting in less than two minutes. Of course it takes practice as does all microscope work.

If your microscope is not equipped with a condenser, then you must simply try by turning the mirror up and down and by changing the stage diaphragm opening to find a good illumination. If you work with high magnifications without a condenser, the concave mirror must be used. Also in microscope work without a condenser, a proper artificial light source is preferable to daylight.

EXAMINING AND POSITIONING PREPARATIONS

Once you have focused your preparation sharply, you naturally want to look it over carefully. The next step is to determine if the iris diaphragm is being used properly for the proposed preparation. Look through the eyepiece and at the same time narrow the diaphragm opening until only a slight aperture remains. If the image becomes considerably sharper and hitherto unseen structures appear, this is a sign that the diaphragm was opened too wide. With each new preparation and each change of objective the diaphragm adjustment must be checked and corrected if necessary.

For many preparations you will make a remarkable discovery at first sight: the microscope delivers a sharp image of only one level of the

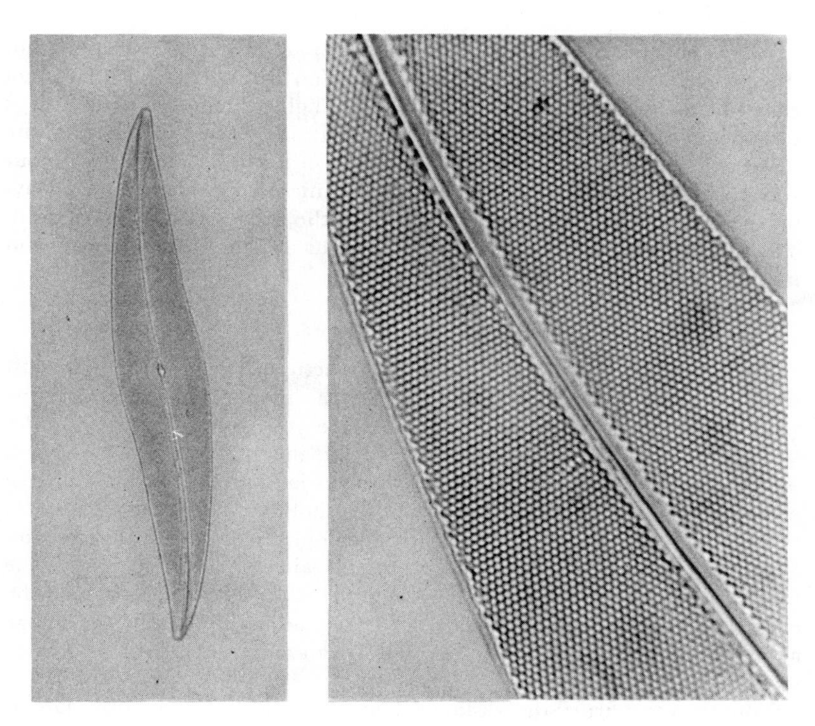

Fig. 13. Diatom shell. (left) Weak magnification (270 times) results in weak detail and wide visual field. (right) Strong magnification (1,825 times), strong objective lens (oil-immersion) results in high detail and narrow visual field.

examined object. This is because the focal depth of a microscope is very small. As a beginner you will attempt to obtain sharp focus by the natural accommodation of your eye. This will work within close limits; but there remains a grave risk because your eye is unusually strained by this. So, for a sharp focus, you must use the micrometer adjustment. Keep your left hand constantly on the micrometer adjustment for all your microscopic examinations. By making small back and forth adjustments, you will focus sharply on one and then another level of the examined object—your eye has nothing further to do but look comfortably and without straining its own lens. The adjusting mechanism of your eye is replaced in microscope work by the operation of this micrometer adjustment.

Now you may want to look at another part of the preparation. For this purpose, move the slide on the stage. You will notice that this is not easy to accomplish at first: even with the slightest movement the image disappears completely from the field, and also moves in the opposite direction to the way you push the slide. In order to be able to find your way through a preparation, look into the microscope—left hand always on the micrometer adjustment—and move the slide by slowly pushing it. You must practice this feat until you can bring any part of the preparation into the field at will.

CHANGING OBJECTIVES

To acquaint yourself with the exact structure of an object, you must make use of a stronger objective. Clamp the preparation tightly on the stage with the spring clips so that the place you wish to examine under stronger magnification cannot slip away. Bring the tube up high, and unscrew the objective—if possible without jarring the microscope. In order to protect the objective from falling, hold your left hand under it; this is really important when the stronger objective is being screwed in. When you have finished, lower the tube again very carefully, until the front lens is a few millimeters away from the slide. Then look through the eyepiece and raise the tube with the coarse adjustment until the image appears in the field. Make the fine adjustment again with the micrometer screw.

You then confirm: the distance between objective lens and slide is much smaller when sharply focused than with a low-powered objective. The diameter of the field which you are able to cover has become smaller. The image seems much darker. The object not only looks considerably larger, but also shows structures which were not recognizable before. (The more powerful objective also has a higher resolution.)

Changing objectives is not too time-consuming, but it often happens that the most interesting place in the preparation is lost by the time a greater magnification is achieved. The process is made considerably simpler it you have attached to the tube a revolving nosepiece which carries three or four objectives. A simple turn of the nosepiece brings each in turn under the tube, so that nothing more than a slight twist is required to make a change in magnification. Because the nosepiece is such a work-saver, especially for the beginner, it is advisable to get one right at the beginning. It is best to have a triple or quadruple nosepiece, even if you have only two sets of objectives. Later you will certainly get an extra one or two objectives. With the triple nosepiece you have the possibility of bringing a weak, a stronger and a still stronger objective into use; with these you will be able—but for exceptional cases—to tackle any work.

LEARNING TO SEE WITH ONE EYE

Look into the microscope and try at the same time to keep the other eye open. At first this might not be possible for you—again and again you will catch yourself trying to close the eye which is not being used because otherwise you "would not see anything". You must teach yourself to reject the images which the free eye gives you. After practice you will always find it easier to keep both eyes open.

Working with one eye closed is extraordinarily tiring. Sustained microscope work and examinations of the most delicate structures are only possible if you keep both eyes open.

If, in spite of the best will-power and practice, you find it impossible to reject the unwanted images from your free eye, then you can use a shield as an emergency measure. Such a shield (shown in Fig. 14) can be easily made by anyone.

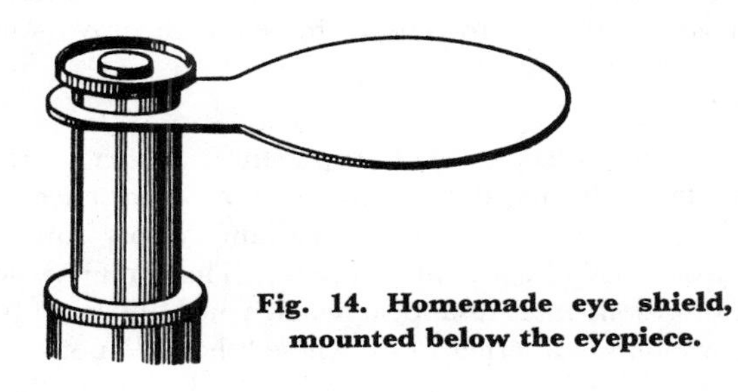

Fig. 14. Homemade eye shield, mounted below the eyepiece.

People who wear eyeglasses will find it best to remove them, especially for long, strenuous work. Only if quick changes between microscope work and other tasks are necessary would it be advisable to continue wearing glasses. Observation with glasses on is very uncomfortable because the surface of the eyeglass lens is constantly scraping against and scratching the eyepiece; after a while the eyeglasses and the eyepiece look as if they were rubbed with sandpaper. This scratching can be avoided by the use of a so-called ocular guard, which anyone can fashion from strong cardboard or hard rubber (see Fig. 15). It can easily be shifted from one eyepiece to another and it will hold even the most strongly curved eyeglass lenses far enough away so that all contact and scratching of the lens surface is impossible.

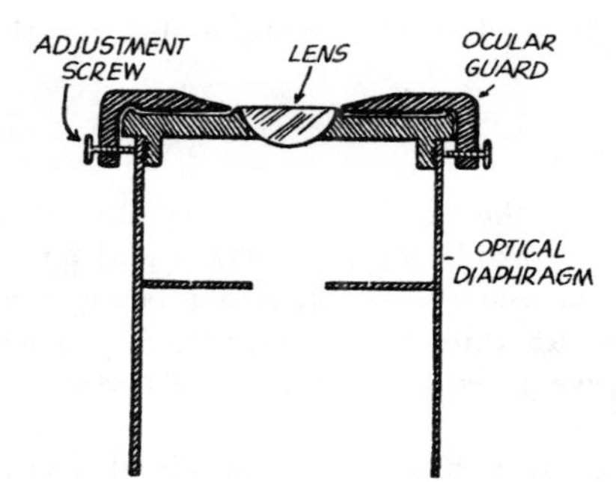

Fig. 15. Ocular guard for eyeglass wearers.

For a start, accustom yourself to looking into the microscope with your left eye. Quite often (as when drawing) your right hand will be needed for other tasks, and the right eye will be needed to observe this hand. Later, when you have gathered a little more experience, you can also train your right eye to relieve the left with drawn-out work.

SEEING "FLIES"

Every microscopist will experience, under certain circumstances, especially with long drawn-out work, a disturbance in his field of vision which causes the appearance of certain sharply outlined bright and dark spots, or rows of spots and threads. These march slowly through your field of vision, then disappear when you blink or rub your eyes, only to appear again in another form. These "flies" do not belong to the preparation, of course, but are an indication of changes in the eyes of the viewer. They are not a sign of illness, but can be very disturbing. In order to prevent such a disturbance, avoid everything which could affect your eyes in any way (smoke, dust, glare, rubbing, irritations, bloodshot eyes, etc.).

DRAWING MICROSCOPIC OBJECTS

You should get in the habit from the beginning of drawing all objects you see through the microscope. It is not at all important to make your drawings "pretty", but it is important for each individual structure to be as correct as you can possibly make it. Drawing things will force you to observe closely, because you must at all times compare the microscopic image with your sketch. You have not truly seen something until you have sketched it, and every microscopist is astonished again and again at how many details he had overlooked until he sketched them. The seemingly harsh proverb that "whoever cannot draw, cannot see" is especially apt in microscopy.

Usually it is tedious and time-consuming to draw all the structures which lie in the visual field. You must, however, avoid any simplifications when recording by sketches. Every sketch of a microscopic preparation should be a faithful copy, not a superficial impression. Also the smallest and most delicate detail of the drawing should agree with the image in the microscope and must be compared constantly and corrected if necessary. In this way you will learn in time to understand all the many phases of microscopically minute forms, and to classify them. In many cases, you will amuse yourself by drawing one or another portion of the field of vision; you might want to isolate about 5 or 10 cells from a multicelled structure.

Fig. 16. Correct position for drawing at the microscope. Look into the microscope with your left eye, holding your right eye open; adjust the preparation closely with your left hand, so that your right hand is in its desired position, free for drawing.

Another bit of advice about drawing microscopic objects: it is good to draw the structures considerably larger than they appear in the visual field. The finer details are then easier to draw.

There are many drawing aids for depicting things exactly for scientific purposes. They are not cheap and the beginner can do without them.

Perhaps at this point a warning could be injected about microphotography. Modern microphotography accomplishes a great deal. However, the preparation necessary to make usable microphotographs is—besides good apparatus—a complete knowledge of microscope technique, especially the methods of lighting. You must first get well acquainted with your microscope, with observing and drawing, before you even think of photographing your preparations.

FIFTEEN WORKING RULES FOR BEGINNERS

1. Always keep your equipment clean and in good working order, so that you are not always looking for something or cleaning it.
2. Never leave your microscope standing on the work table after finishing an examination; it will become dusty. Always replace it in its case or put a plastic cover over it.
3. When moving your microscope never pick it up by the part which has the micrometer adjustment. Use the limb for this purpose.
4. Avoid turning the micrometer adjustment too far. It can only raise or lower the tube by a few millimeters. When these limits have been reached, it should be turned back to a middle adjustment, and the coarse adjustment used first.
5. Never adjust the tube from a high to a low position, but rather from a low to a high one; otherwise—especially with high-power objectives—you may carelessly crush your preparation and damage your objective.
6. Never use direct sunlight for microscope work. In the first place it will hurt your eyes, in the second place the flood of light streaming in will erase all fine details, in the third place the bright light will damage the mirror and the enamel with which the instrument is coated, and lastly you will cook your specimen! Diffused light is best for this work.
7. Never close one eye while you look in the microscope with the other.
8. Never work in a cold room. Here the lenses will take on vapor from your breath, fogging the image.
9. When working with reagents, be careful not to get any on the front lens. In many cases (strong acids and alkalis) the objective can be rendered useless if this happens. In any case, the result will be unclear images. Avoid using cover-glasses which are too small.
10. Never allow reagents to stand uncovered near the microscope,

especially acids. By the same token, the microscope should never be stored in a cabinet with reagents. Sooner or later the microscope will suffer as a result, because even the vapors of some reagents cause damage to metals and glass.

11. Never unscrew the objective lenses, and unscrew the eyepiece lenses only when they have to be cleaned. When objective lenses need cleaning, which is seldom the case, they should be returned to the manufacturer for this work.

12. Never use alcohol or xylol for cleaning the front lens of the objective system, because it could result in fogging (caused by cement melting). For cleaning, use an often-washed linen cloth, dampened with a little distilled water. If the lenses are smudged with oil, then the cloth should be dampened with a little benzine.

13. Never allow alcohol to touch the enamel or lacquer of the microscope. Use distilled water or a little benzine, applied with a soft brush and a soft linen cloth.

14. Never try to loosen a micrometer screw yourself. For this, the instrument must be sent back to the manufacturer.

15. Do not use oil to lubricate the coarse adjusting screw or the tube carrier. It promotes an accumulation of dust and becomes dirty when it gets thicker. Clean vaseline, wiped with a soft linen cloth to an extremely thin coating, is a good lubricant for microscopes.

3. Examining Simple Preparations

Examining fresh material without previous preparation always takes precedence in microscopy. Even later, when you have learned to prepare "permanent preparations" (preparations which can be saved up for decades) you will work with the fresh material, if possible, at the beginning of each examination.

TABLE-SALT CRYSTALS

Place on a slide a small drop of water with the aid of a pipette and place in it a fragment of table-salt the size of a pinhead. Naturally the water dissolves the salt. Leave the solution in a dust-free place and allow it to evaporate. There will remain on the slide a little white patch. Cover this with a cover-glass—only in rare cases is an examination made without a cover-glass—and then observe this simple preparation under the microscope. As always, first adjust the weakest objective lens. You can plainly see smaller and larger squares with dark edges and parallel lines. The center is usually darker also and the diagonal lines appear very plainly: these are table-salt crystals (Fig. 17).

Place a very pretty, large crystal in the center of the visual field. With the micrometer adjustment, raise and lower the tube slightly and observe in this way the difference in clarity at different levels.

You can add another interesting experiment here: remove the microscope mirror, without changing anything else in the adjustment. Looking into the microscope now, note that the table-salt crystals take on a beautiful three-dimensional appearance on a dark background. They are now illuminated from above (toplighting, or "opaque lighting").

Draw the observed images exactly. For this task the drawing card (Fig. 10) is very convenient.

Replace the mirror and observe again with transmitted light. Now place a drop of water at the edge of the cover-glass. (Don't get any on top of the cover-glass!) As soon as the water is drawn under the cover-

Fig. 17. Table salt crystals, magnified 125 times.

glass, look quickly into the microscope: the crystals will become more transparent, lighter, and the corners and edges will become rounded. Soon the crystals are dissolved. Now dark, sharply outlined circles appear: these are air bubbles.

AIR BUBBLES

Focus on one of the larger bubbles, observe it closely and then hold your hand over the mirror, so that only toplight falls upon the bubble:

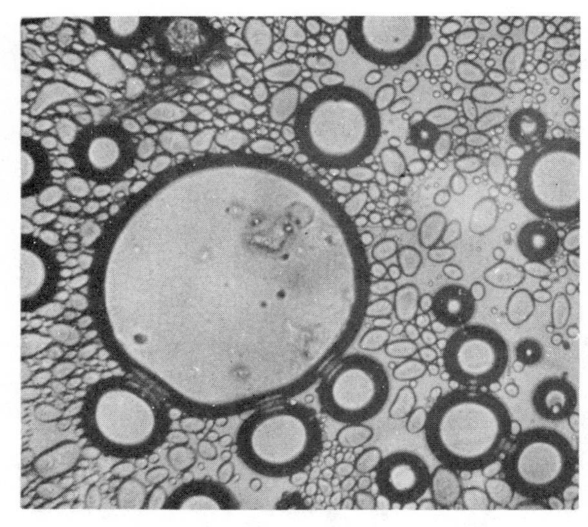

Fig. 18. Air bubbles.

the air bubble appears as a brightly gleaming ring on a dark background. You must remember the appearance of air bubbles, because beginners especially confuse air bubbles with structures of the object being examined. In case of doubt, the following test is useful: simply hold your hand over the mirror and look at the object with toplighting; the characteristic rings are produced only by air bubbles.

You must accustom yourself to cleaning all equipment immediately after use. The parts of the microscope on which your breath has fallen should be wiped with a soft linen cloth.

POTATO STARCH

Cut through a raw potato, scrape a little from the freshly-cut surface into a drop of water on a slide and stir it well with a needle. Then put the cover-glass in place.

For this, a little practice is advisable. Grasp the lightest coverslip by the edges between your thumb and index finger and rest one edge on the slide so that it forms an angle of 30° to 40°. Now push the coverslip towards the drop until the coverslip edge resting on the slide touches the water; not until then do you let the coverslip sink slowly onto the slide, until it rests flat. By letting it down slowly you will avoid air bubbles which would interfere with examination.

Probably at first you will not be fortunate enough to measure the drop properly. If you use too much water, the coverslip will float, and you must sponge up the surplus water with a piece of filter paper or blotting paper; if the drop is too small, then you can see many small and large, irregular bubbles. In either case you cannot do much with the preparation. It is better to practice beforehand to judge the size of the drop of water: the drop should be just big enough to fill the space under the coverslip when it is properly in place.

Now your potato preparation is ready. With a weak objective you will see tiny rounded, elliptical bodies. The stronger objective shows that the borders of these little bodies are of irregular shape; inside of each grain you will find, when you diminish the light (by means of the iris diaphragm), a delicate arrangement of layers (Fig. 20). All these little bodies are grains of starch. Starch, an important reserve material of the potato and many other plants, is accumulated in special storage organs—especially the tuber of the potato. Such storage organs of plants are used by man for his own nourishment.

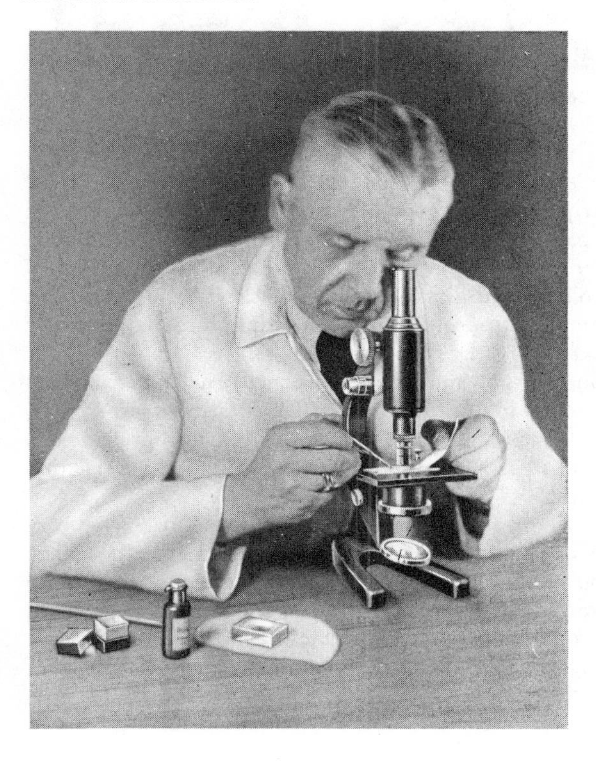

Fig. 19. Drawing a fluid under the coverslip.

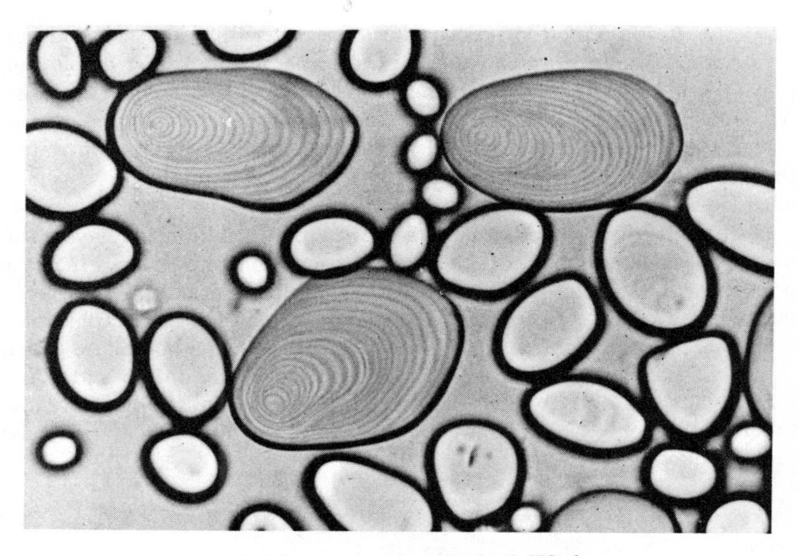

Fig. 20. Potato starch, enlarged 450 times.

Now place on the edge of the coverslip a small drop of tincture of iodine and hold a strip of filter paper against the opposite edge of the coverslip. The filter paper will absorb some water, allowing the iodine to make its way under the coverslip. You now find that all the grains of starch that come in contact with the iodine take on an intense blue color. Within a few moments (or at once, if too much iodine is used), the color turns to a deep black.

This iodine reaction of starch must be remembered: in later examinations you will often come in contact with starch grains of every sort, whether in plant cells, or in nutritive substances, or as impurities in microscopic preparations. In all doubtful cases, the iodine reaction provides an infallible test.

When you work with reagents containing iodine you must be very careful that the iodine does not touch any metal parts that hold the lenses. Not the slightest drop of iodine must touch the microscope, and at no time may such an examination be made without a coverslip. Slides and coverslips that have come in contact with tincture of iodine can never be used again; even the slightest traces of iodine will disturb the coloring of permanent preparations.

If you inadvertently get traces of iodine on a slide or cover-glass and then use it for a dyed permanent preparation, your time—often hours and days of it—is wasted: even after a short time the preparation will become pale. Therefore throw away iodine-stained slides and cover-glasses, or wash them thoroughly in a sodium thiosulphate solution.

For another preparation of potato starch, use a drop of 2 to 4% caustic potash solution. The grains of starch will then become larger, swell up and eventually dissolve.

If you examine the contents of grains of wheat, oats, barley, corn or rice in the same manner, you will also find a great many grains of starch. The shape of the starch grains is totally different, depending on the type of plant. You can in this way often identify flour and other starchy foods by a simple microscopic examination. Further valuable objects for starch examination are the legumes (beans, peas and lentils).

ONION SKIN (*Allium cepa*)

Cut an onion the long way through the middle. Doing this, note that the onion is composed of separate layers which are closely fitted into each other. Separate one of these layers and with your tweezers peel off from the inside, concave surface a bit of the top skin. Cut out a small square of this skin and place it on a slide in a drop of water.

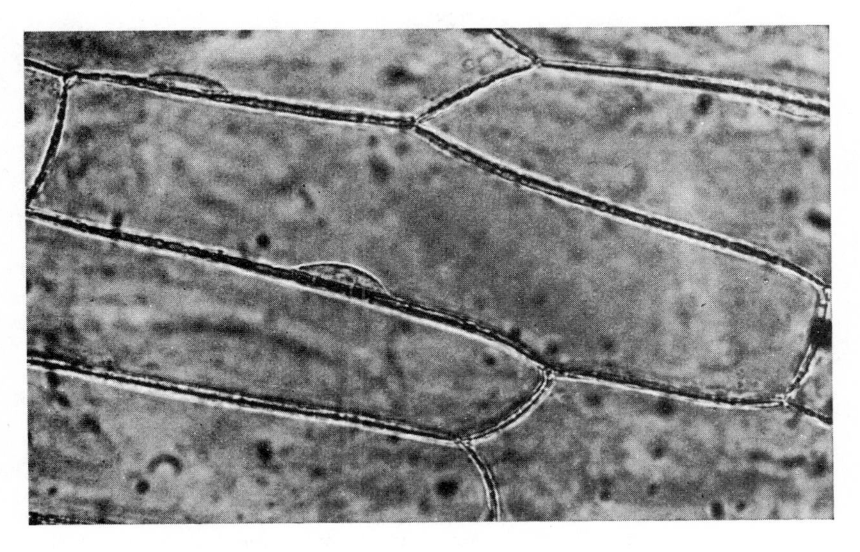

Fig. 21. Outer-skin cells of onion skin.

After putting on a coverslip you will see under medium magnification elongated, often six-sided cells (Fig. 21). With stronger magnification you can make out inside each cell a delicate little bubble which usually lies near the cell wall. This is the cell nucleus, which you will soon see more clearly.

Introduce into the preparation a drop of tincture of iodine. The cell nuclei first become yellowish, then a deep brownish yellow; the nuclear bodies (nucleoli) found in the cell nuclei appear after iodine treatment as sharply delineated little circles. If the preparation is successful, then the yellowish-tinted, finely-grained cytoplasm can also be seen. The cell is not completely filled with cytoplasm; most of the cytoplasm lies as a thin layer along the cell wall. Mature plant cells contain one or more large, sapfilled cavities called cell sap cavities or vacuoles.

PLASMOLYSIS

Make a fresh preparation of onion skin and draw a drop of glycerin under the coverslip. The glycerin draws water from the cells, shrinking the cytoplasm and causing it to fall away from the cell wall. Eventually it forms a more or less globular mass inside the cell (Fig. 22). This proce-

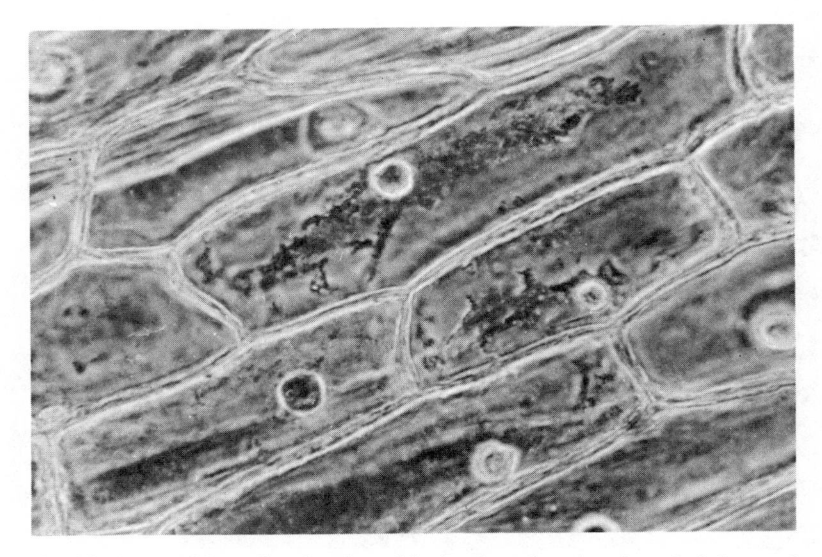

Fig. 22. Outer-skin cells of onion skin in glycerin, showing shrinkage of cell contents.

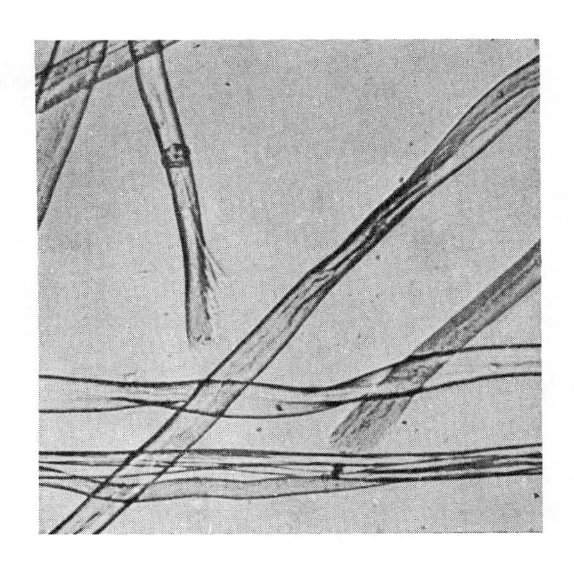

Fig. 23. Cotton.

dure is known as plasmolysis. If plasmolysis is interrupted at the right moment by placing the skin again in clean water, this shrinking procedure can be reversed (deplasmolysis). Plasmolysis is brought about more perfectly if, instead of glycerin, you employ a heavy sugar solution.

COTTON

Place several cotton fibers (cleaned cotton-wool) under the microscope. Each individual strand is a seed hair of the cotton plant (*Gossypium*) and consists of only a single cell. Notice the cell walls and particularly a unique quality about cotton that distinguishes it from other textile fibers: in many cases the cell fibers are twisted (Fig. 23).

NETTLE HAIRS

From the young leaves of the stinging-nettle (*Urtico dioica*) cut away some of the hairs with a razor and place them in water. Work carefully here because the nodes of the hairs break off easily. Each hair bellies out at the lower part and sits in a many-celled structure (Fig. 24). The point of the hair is bent and ends in a round knob. At the lightest touch this little knob breaks off at an angle. In the cell walls of the hair there is stored silicic acid and calcium carbonate, which make the hairs stiff and brittle. Draw some hydrochloric acid under the coverslip and the lime

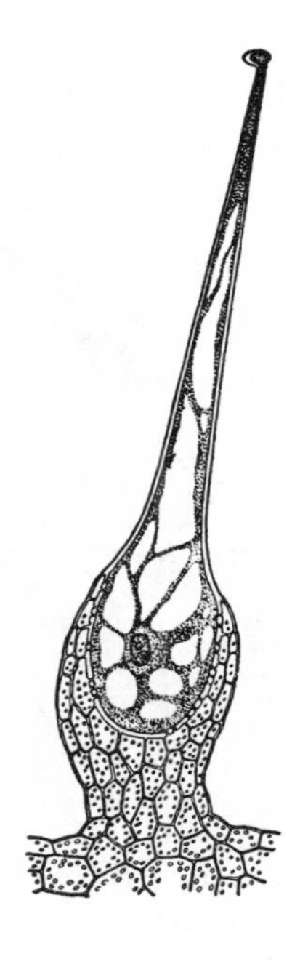

Fig. 24. Stinging-nettle hair.

will dissolve and foam up. (Care must be taken not to let the hydrochloric acid touch any part of the microscope.)

CANADIAN WATERWEED (*Elodea canadensis*)

Cytoplasmic streaming. At an aquarium dealer's, or from a nearby pond, get a few strands of Canadian waterweed. Tear off some young leaves with your tweezers and examine them in water. This is a little more difficult object to examine than those previously described: for example, only a single layer of cells of onion skin was used, but with the waterweed leaf there are two. As the microscope can show only one layer of an object with clarity, you must therefore work busily with your micrometer adjustment.

First you will find rectangular cells, in which there are a great many green discs (Fig. 25). The green discs are chloroplasts, which give the plant its green color. Now focus with a medium to strong magnification on some cells near the center rib and observe. With luck, you will see in a few minutes that the chloroplasts seem to move; they wander in one

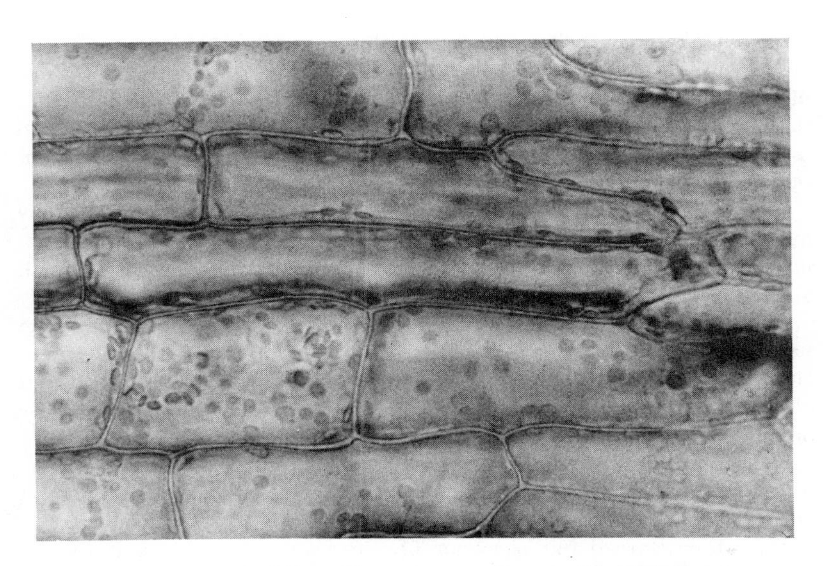

Fig. 25. Leaf of Canadian waterweed, magnified 450 times.

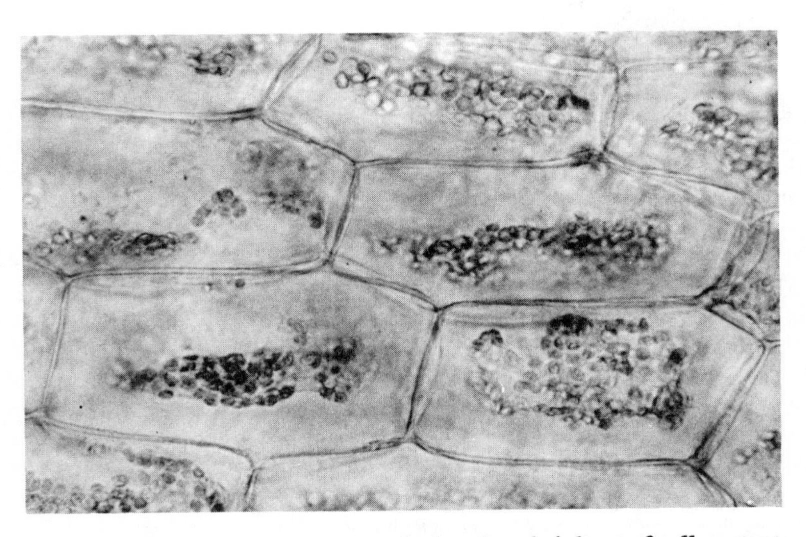

Fig. 26. Leaf of Canadian waterweed, showing shrinkage of cell contents after addition of glycerin.

direction along the cell walls and in this way travel around the entire cell. However, we know that the chloroplasts are in themselves incapable of motion—so it must be the cytoplasm in which the chloroplasts are suspended that gives them motion. The cytoplasm cannot be seen in an untreated, living preparation; you can conclude, however, by the apparent movement of the chloroplasts that it flows.

The cytoplasmic stream does not always go into motion within a few minutes. Sometimes it takes up to a half an hour, and in rare cases the flow cannot be seen at all. You must then repeat your observations with a leaf of another stalk.

If you examine older leaves with larger cells in the same way, you can see the large central cell sap space left by the chloroplasts, which remain close to the cell walls. Add a drop of glycerin to such an older leaf and observe the plasmolysis (Fig. 26), which you saw more clearly in the onion skin.

Take a fresh preparation and this time draw alcohol under the coverslip: the cytoplasmic streaming comes to a complete standstill almost at once. Alcohol kills the cell contents.

Stem Apex. Spread out a sprout of waterweed on a black-and-white preparation plate[3], on whichever part is best suited for the work; then

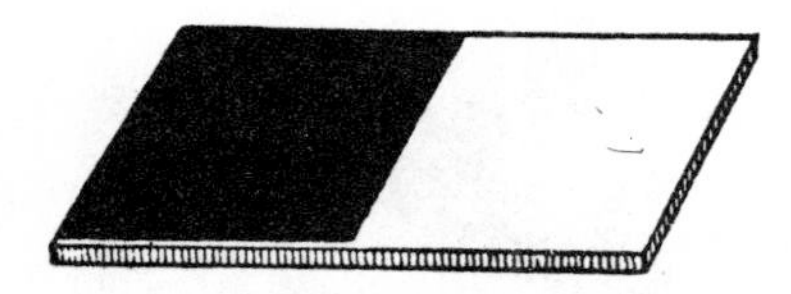

Fig. 27. Black-and-white preparation glass.

douse it with water. With two preparation needles carefully remove the tiny leaves which form a bud, until the stem apex appears (Fig. 28). You may need a magnifying glass for this work. Isolate the stem apex and examine it with weak magnification: growing point, leaf primordia (rudiment) and the arrangement of the cells.

SPURGE (*Euphorbia*)

Break off a stem of this well-known herb, and catch on a slide one of the drops of milky fluid which pours out. Cover it with a coverslip. Under the microscope you will see a great many dots, and here and there some larger rod-shaped bodies (Fig. 29). If you allow a drop to dry, then you

[3] If you do not wish to purchase such a preparation plate, you can easily make one (Fig. 27). Take a piece of glass of sufficient size and paste on its underside two pieces of paper, one black and one white, each half the size of the glass, making a two-part backing. Depending upon the nature of your objects you can place them either on the black or the white part.

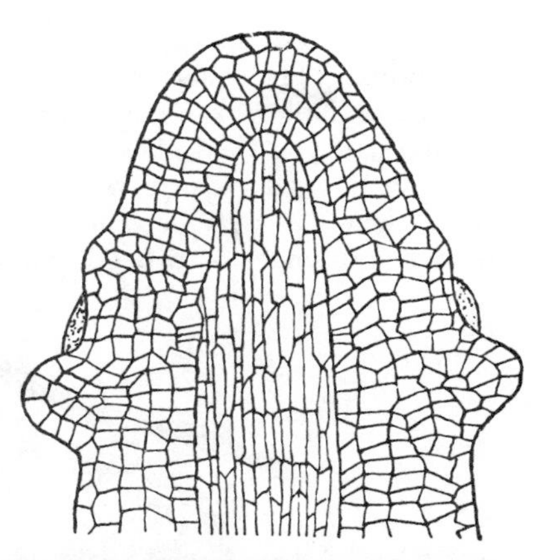

Fig. 28. Growing point of Canadian waterweed. Schematic.

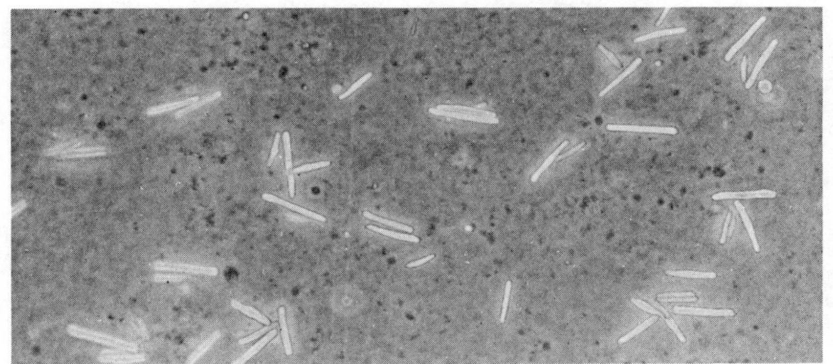

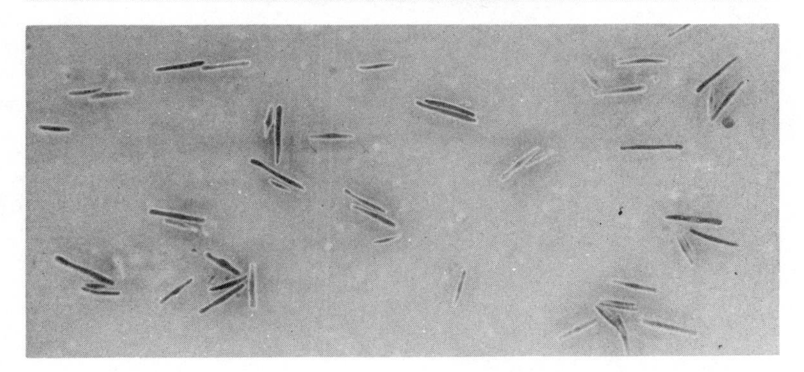

Fig. 29. Starch rods from a drop of *Euphorbia* sap. (above) Undyed. (below) After addition of iodine solution, magnified 300 times.

will find the rods more easily because they will seem brighter in their surroundings. Then place a drop of tincture of iodine at the edge of the coverslip and allow it to flow in: the rods turn blue, some parts becoming very dark. If you then place some caustic potash solution at the edge of the coverslip and blot up the moisture from the other side with filter paper, then the rods become light again, expand somewhat, become broader and longer, and bend. These are starch crystals.

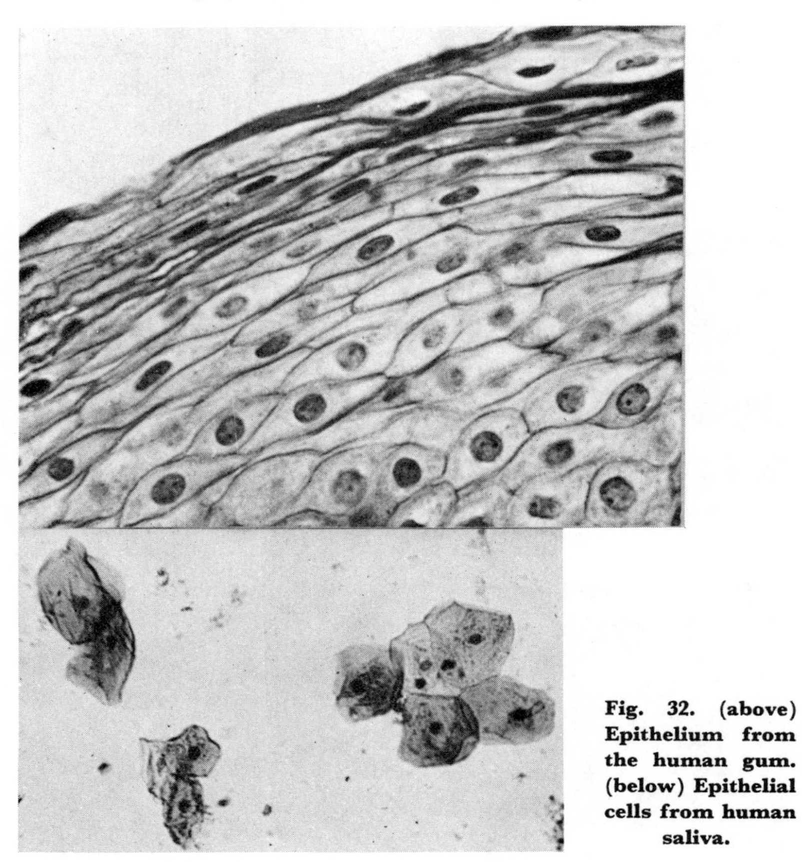

Fig. 32. (above) Epithelium from the human gum. (below) Epithelial cells from human saliva.

CELLS FROM INSIDE THE HUMAN CHEEK

You have already become acquainted with plant cells by examining onion skin and the leaf of waterweed. However, the cells of animals and humans are totally different. Of course, you will also find cytoplasm and cell nuclei as in all living cells—but the outer shape of the cells of animals and humans is much more changeable, more irregular, and far less rigid than plant cells. Plant cells are enclosed in firm membranes, the cell walls, which form the unchangeable structure of the individual plant cell. Animal cells, on the other hand, are not surrounded by rigid membranes —a thin wisp of firmer plasma surrounds animal cells and keeps them separate.

Take half of a wooden matchstick or other similar piece of wood and scrape off a little material from inside your cheek. Put it on a slide. If it is too thick, thin it with a drop of saliva.

With the help of medium to strong magnification you will see flat, irregular structures with rounded corners; these are epithelial cells, that is, cells covering the outer tissues (Fig. 32). You will find these epithelial cells still partly connected with each other and partly loosened from the tissues. If you cut down the light by closing the iris diaphragm, you can also make out cell nuclei as rounded or elongated little bubbles inside the cells. Often there can be found on the surface of the cells highly opaque particles and rods: these are bacteria, whose closer acquaintance you will make a little later.

Usually there are a great many other ingredients to be found in these scrapings besides epithelial cells and bacteria. If you observe the smallest particles, you will see that they are in constant quivering motion. This Brownian movement is not a sign of life; on the contrary, it depends on the heat movements of the surrounding molecules. The molecular motion can be studied better if you add a drop of ink thinned with water. The ink particles dance around aimlessly in the visual field.

The cell nuclei of the epithelial cells can be made more visible if you introduce a drop of Löffler's methylene blue (a dye which you will need later for bacterial study in any case) under the coverslip. The nuclei will become dyed a deep blue in a short time.

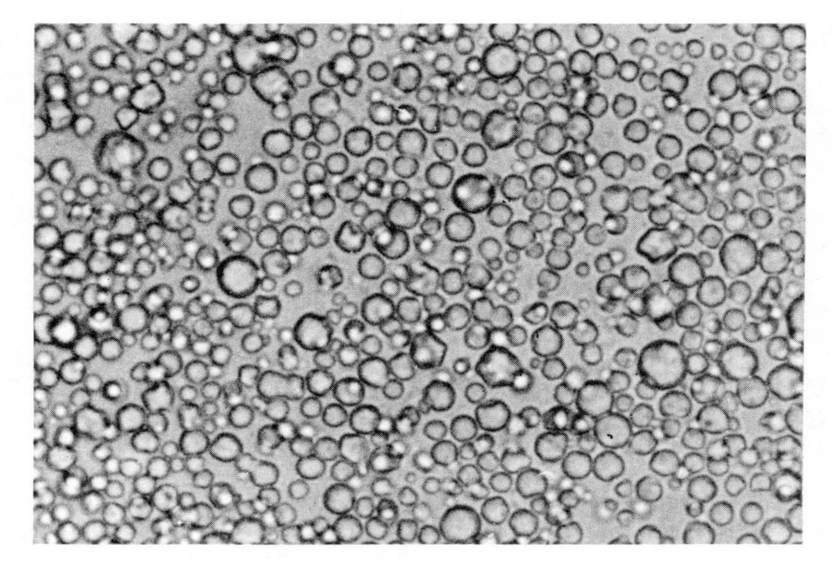

Fig. 33. Fat globules in milk.

MILK

Microscopic examination of a drop of milk will show you countless globules of fat in varied sizes in a clear liquid (Fig. 33). It is because of the number of these fat globules that milk loses its transparency when in thicker layers.

BLOOD

You will require a 10% salt solution (10 gm. of salt placed in a beaker with 100 gm. of distilled water); an 0.9% salt solution; slide; coverslip scrupulously cleaned with alcohol; and an unused sharp pen-point. If a sensitive scale or balance cannot be obtained, the 0.9% salt solution can be thinned out from the 10% solution: in the graduated glass place 9 cc. of the 10% solution and then fill it to 100 cc. with the distilled water.[4]

You cannot simply take a drop of blood, put it on a slide and then examine it microscopically. Each drop of blood contains such a tremendous number of blood cells (corpuscles) that in the thicker layers nothing could be seen. You must either spread out the blood in a very thin layer, or thin it considerably. These are the two methods:

[4] This solution can be purchased as "saline" from any pharmacy.

Clean off the tip of your little finger with a strong alcohol solution and then prick it with the pen-point, which has first been heated and then allowed to cool. Wipe off the first drop of blood, and let the second fall into a watch-glass which has been filled with 0.9% salt solution. Stir it immediately. With the third drop, finally, make a "smear" preparation.

The preliminary to a smear preparation is simple, but requires fast and neat work. The drop of blood to be used must be absolutely fresh; after only a few seconds out of the wound it is unfit for smearing. Place the drop on the very end of a scrupulously clean slide. Place a second slide which is just as clean so that it forms an acute angle to the other. Now push the upper slide so that the drop of blood lies in the angle formed by both slides (Fig. 34). As soon as the blood touches the upper slide, it spreads all along its edge. Now if you push the upper slide slowly and evenly, the spread-out drop pulls along and forms a broad, very thin layer of blood behind it. Such a layer should have a pale yellowish appearance. If it shows a reddish shimmer, it is too thick.

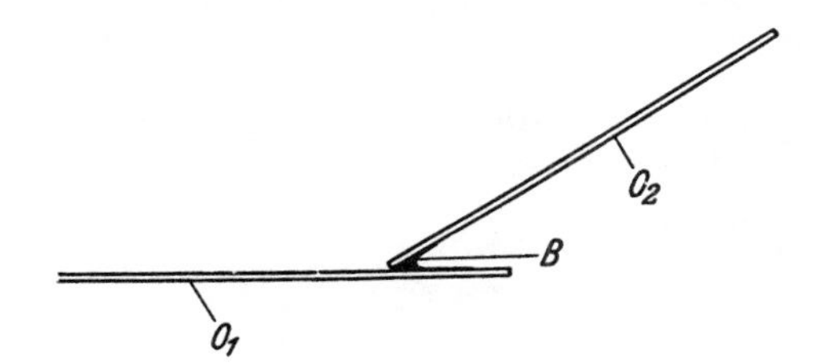

Fig. 34. Preparation of a blood smear. B is a drop of blood to be spread. O_1 and O_2 are slides.

The microscopic examination of the smear will show you many circular discs, the red blood cells. In the center of each blood cell you will see a lighter or a darker shadow. In a good smear, each blood cell should lie by itself, without covering others or touching them (Figs. 35, 36).

Such smears are very valuable for closer examination. They have a disadvantage, however: you can only see one side of the blood cells and cannot even imagine a side view. Here you will now be helped by the second method of thinning. Use the 0.9% salt solution into which you diluted the drop of blood previously. Place a drop of the dilution on the slide, place a coverslip on it and examine it. The blood cells are pulled along in the stream which every fluid preparation has at first. Sooner or later one or the other will turn over, and at this moment you can observe the view from the edge. You will see that the red cells observed from the side are not—as you might have been led to believe by the smear preparation—lentil-shaped in appearance, but rather dumbbell-shaped. They are

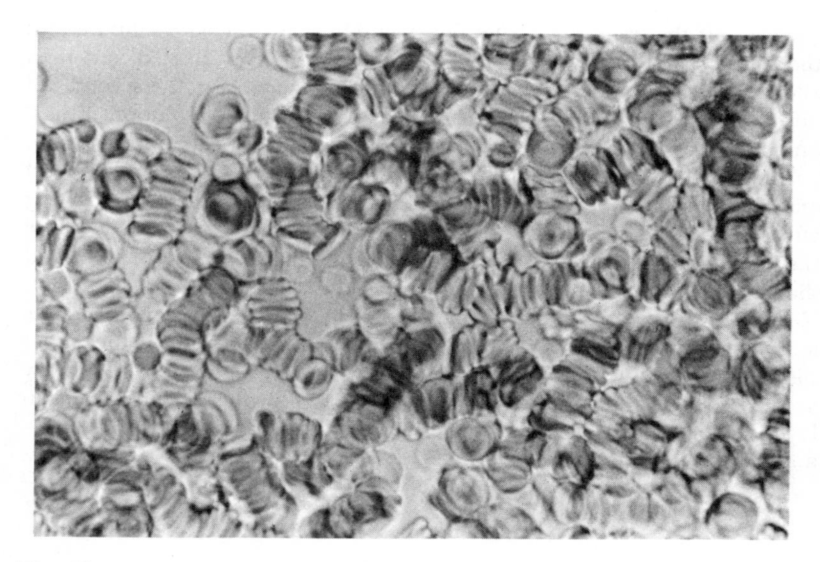

Fig. 35. Human blood, undyed. This preparation is too thick; the red blood cells are crowded together like rolls of coins.

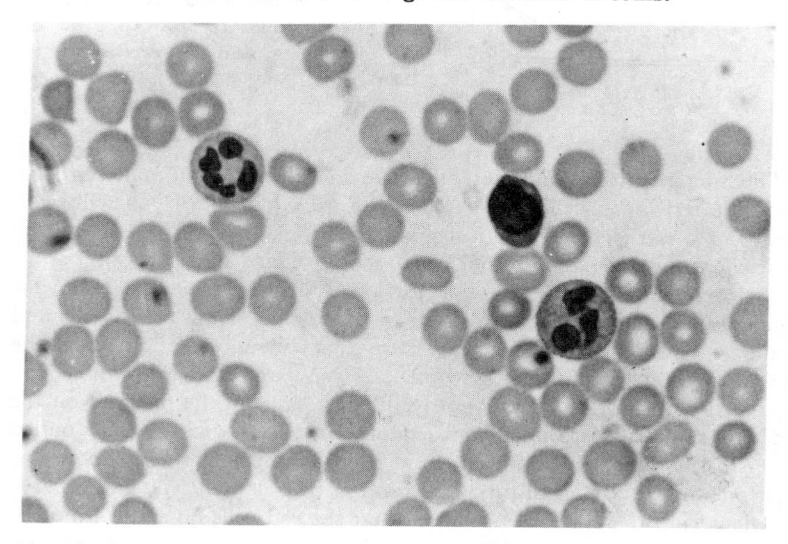

Fig. 36. Human blood, thin smear, dyed. Besides the red blood cells three white blood cells may be seen with cell nuclei.

indented in the middle on both sides, and these indentations are what you saw as shadows in the smear preparation.

Now draw a 3% solution of acetic acid under the coverslip. The red cells become lighter and finally disappear altogether. The acetic acid dissolves the blood cells. In your preparation you will now see many tiny grains—the remains of the blood cells—and a very few round, colorless bodies. In these colorless bodies you will see a small, lobed or round inner body, the cell nucleus. These are white blood cells which, unlike the red ones, have an additional cell nucleus; they are rarer in the blood than the red cells. (One cc. of blood contains about 4 to 5 million red cells, but only about 8,000 white ones.)

Take another preparation and draw distilled water instead of acetic acid under the coverslip, and with a third preparation use a 10% salt solution. You will find that the red cells burst in distilled water, but in the 10% salt solution they shrink and take on a spiny appearance. The salt concentration inside the red cells amounts to about a 0.9% salt solution. In this type of solution, which for this reason is called "physiological saline", the form of the blood cells is maintained. The 10% salt solution draws water from the blood cells and thereby causes them to shrink.

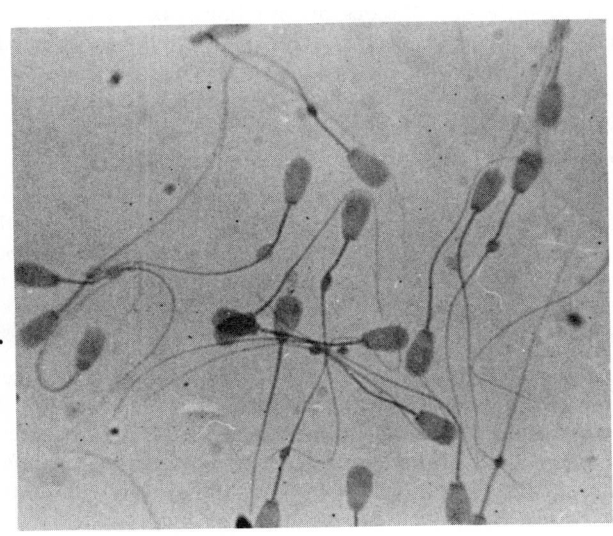

Fig. 37. Ox sperm greatly enlarged.

SPERM CELLS (*Spermatozoa*)

These are handled in a similar manner to blood. Cut the wall of a sperm-gland from a rabbit or a freshly-killed ox or steer (you can probably get one from a butcher, or, if not, from a slaughterhouse) and allow a drop of the fluid which flows out to mix into the physiological salt solution. Observe: head, body and thread-like tail (Fig. 37). Motion: when the fluid is drawn from one side notice that they swim against the current.

HUMAN HAIR

Examination is made in a drop of glycerin. You will see the three components of hair (Fig. 38). In the center is the medulla, which builds up a rope of cells, one following another; the medulla, especially in the hair of older persons, contains many air bubbles, causing the hair to appear white. Surrounding the medulla is the cortical substance (cortex). This main component of hair consists of horny epidermal (outer skin) cells and contains the pigment grains which give the hair its color. Further outside, the hair is completed by the cuticle (outer skin), which is composed of flat, grainless, scaly-layered cells.

Freshly plucked hair shows a medullary cylinder of dark stripes when examined in water. If you allow a drop of glycerin to flow on to a dry hair under the microscope, you can readily see the air being pushed out of the medullary cylinder. Every air bubble in the medulla appears as a silvery little dot. Besides human hair, other hairs are recommended as material: sheep's hair (wool), which often has no medulla visible; the hair of a bat, with cone-shaped segments; and hairs from a mouse and a mole.

FEATHERS

Down from young pigeons or chicks, and the tips of covert feathers are handled in a similar manner to hair (Fig. 39). Observe on covert feathers: branches of the first and second order (barbs and barbules), and the little hooks of the barbicels (hamuli).

CLEANING OF SLIDES AND COVERSLIPS

Used slides and coverslips should be placed in a sealable container (for example, an old jam jar) containing industrial spirit (95% alcohol). When they are needed, dry them with a cloth and they are then ready for use. The spirit must be changed from time to time.

For preparations which are sealed in synthetic resin or Canada balsam (you will encounter many of these in the experiments that follow), you

Fig. 38. Human hair.

must take other measures. First, heat the slide over a small flame, causing the sealing material to become so soft that it can be separated easily from the slide. Then place the slides in used xylol or kerosene, which will dissolve the balsam. To remove the xylol or the kerosene, rinse the glasses with fuel alcohol.

Very dirty slides often need to be cleaned with a cleanser and steel wool.

Fig. 39. Starling feather, magnified 120 times.

4. Insect Preparations

The multiform and extraordinary beautiful world of the insect gives the beginner an endless source of educational microscopic preparations, and making them requires no particular experience.

The unusual mouth parts of the various insect species could alone compose an entire collection; then there are the eyes with their angular facets, the antennae, the different kinds of palpae, the legs with their many different kinds of terminal segments, the wings and other parts—all worthy objects which can be prepared without much trouble. The examples chosen will serve as preparations for other similar work.

KILLING WITH CHLOROFORM

The insects should be killed for this purpose with potassium cyanide or chloroform. Potassium cyanide is a dangerous poison and for this reason is difficult to obtain. Chloroform serves fully as well. Obtain a suitable empty tube which can be sealed with a cork (a large pill vial or test tube). Pierce the cork with a piece of wire; the end of the wire should protrude about 2 to 3 cm. from the cork. Wind the wire tightly around a wad of cotton and soak the wad well with chloroform. Insert the cork so that the wad of cotton does not touch the inside wall of the tube. After having caught your insect, remove the cork, imprison the insect in the tube with the chloroform-filled cotton wad (with a little practice this becomes easy) and quickly replace the cork. The chloroform vapors will kill the creature in a very short time.

BUTTERFLY SCALES

Cut a ring of thin cardboard, leaving an inside diameter somewhat smaller than the width of the slides you are using. Dip the cardboard ring in synthetic resin or Canada balsam and—after the excess mounting medium has dripped off—press it tightly against a slide. Over the upper side of the ring apply an additional very thin layer of resin. Now place a piece of butterfly wing on a clean coverslip, in order to get a layer of butterfly "dust" to adhere to the glass surface. You must carefully turn

Fig. 40. Butterfly scales.

over the prepared slide so that the ring faces down, and lay it on the dusty surface of the coverslip. The mounting medium you have applied will stick the coverslip edges firmly together. In the middle you will have a "window" (the ring) through which you can observe the dust from the butterfly wing.

The dust consists of extremely regular scales with horizontal and vertical markings, with toothed or flat edges and varied coloration (Fig. 40). It is intriguing to compare microscopically the scales of different butterflies. You can do this now, because your preparation is a permanent one, which can be kept as long as you like. Be sure to label it.

BUTTERFLY WINGS

These can be mounted whole or cut into suitable pieces. Because a dry preparation would be too thick to let the light through, it is better to make a mounted preparation. Take the wing or a piece which has been cut off with a pair of scissors and place it in a watch-glass or "salt-shaker". Then pour xylol over it. Cover it well, so that the xylol does not become cloudy, and let the whole stand for 2 or 3 hours. The watch-glass can be set on a little watch-glass dish, a rectangular piece of wood with a hole in the center, which can be easily made from cigar-box wood with the help of a fretsaw (Fig. 41).

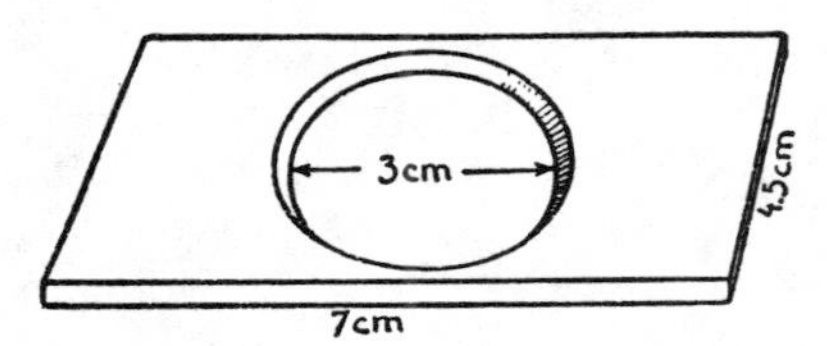

Fig. 41. Watch-glass stand.

Place on the slide a large enough drop of synthetic resin to fit the coverslip and then place the piece of butterfly wing soaked in xylol on it with a needle. With the help of two needles moistened with a drop of xylol arrange the wing so that the preparation forms a clearly arranged picture. Then put the coverslip on, and the permanent preparation is complete. If any air bubbles have crept in, they can be moved to the side with a little gentle heat from a small flame. Then, as soon as they reach the edge of the coverslip, prick them with a hot needle. If, because of thick objects the coverslip will not lie flat, then shim (wedge) the sides with coverslip splinters and fill the open spaces with synthetic resin.

Set the preparation aside, flat, to dry in a dust-free spot. Later, identify the objects permanently with paper or cardboard labels pasted to the left and right. Note on the left label the systematic identification of the object (the common name as well as the scientific name), and on the right label the mounting material, place found, date of preparation, and name of the person making the preparation (Fig. 42).

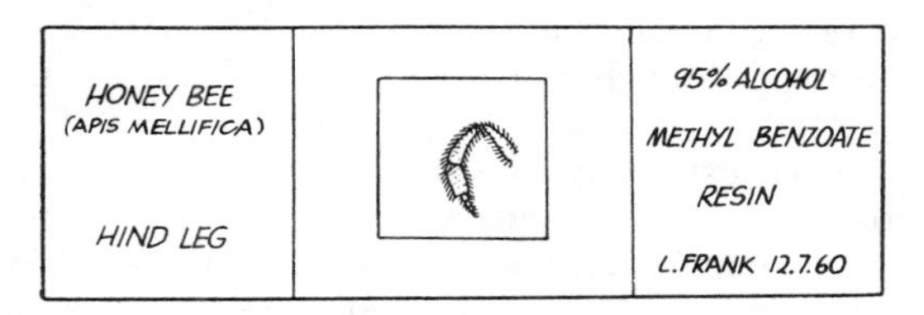

Fig. 42. Correctly mounted and labelled permanent preparation.

Because a permanent microscopic preparation serves its full purpose only when properly labeled, and when it carries all the necessary information for identifying the object fully, every microscopist should make it a habit from the very beginning not to put a preparation into his collection which carries incomplete information.

MOUTH OF JUNE BUG

Kill a June bug with alcohol, ether or chloroform, then place it in a 70% alcohol solution. Work on it as soon as possible. Separate the mouth parts with your tweezers and fine scissors and then place them in a dish of 95% alcohol (industrial spirit) and leave them in this for at least an hour. Then place them in methyl benzoate, which draws out the last traces of water from the tissues and makes them translucent (it "clears" them). The objects must remain in the methyl benzoate until they have sunk to the bottom and appear transparent. In some circumstances this may take hours. You can—if you wish to get on with other work—leave the objects for several days in the methyl benzoate, but the jar of methyl benzoate must be sealed tightly.[5]

Finally, mount the pale pieces in synthetic resin (or Canada balsam). Place a drop of the mounting medium on the slide, and then place the parts on the slide carefully with a needle and tweezers. With two needles arrange the separate parts so that the preparation shows an overall view of the mouth parts of the June bug: upper lip (labrum), two strongly-toothed upper jaws (mandibles), two lower jaws (maxillae) with feelers (palps), and a hinged lower lip (labium) (Fig. 43). Finally place a cover-slip on it and the permanent preparation is finished.

You can treat other insects in the same way as the June bug. Information about the different structures of the mouth parts of other insects is given in any good textbook of zoology or entomology.

If methyl benzoate is not to be had, you can also use oil of cloves or xylol as a clearing agent. Xylol, however, does not absorb any traces of water and for this reason the objects must always be dehydrated before they are placed in xylol. Dehydration by drying out is impossible. The objects shrink so much when dried out that the finer structures can no longer be observed. For this reason you must use absolute (100%) isopropyl alcohol.[6] Take the mouth parts (or similar objects) from the 95% alcohol and place them in isopropyl alcohol; let them remain in it up to several hours, after which you must move them to the xylol where they

[5] Best suited for the purpose are small preparation jars with corks.
[6] Isopropyl alcohol is much cheaper than absolute ethyl alcohol and just as good for this use.

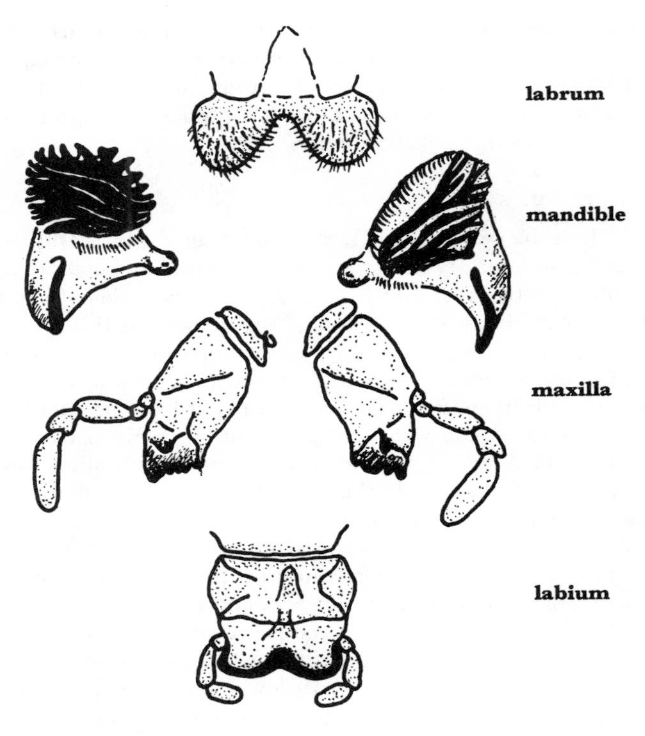

Fig. 43. Dissected mouth parts of a June bug.

remain for hours to days. Then mount them (as described) in synthetic resin.

The synthetic resin mounting mediums (such as Permount) dry gradually in the air. Preparations made with synthetic resins are very durable and you will hardly ever—if the work has been done carefully—have a preparation spoil. The preparations dry relatively slowly. You can speed up this drying process by placing the preparations in a warm place (incubator or in the proximity of a stove). The temperature should never exceed 120° F. It is best to choose a drop of mounting medium which is just large enough to fill the space under the cover-glass. The preparations will then set faster. With cut preparations as well, you should use only as much synthetic resin as is absolutely necessary.

Formerly the sealing material which was most used was Canada balsam, which can be applied in the same way as a synthetic resin. Canada balsam has a disadvantage in that dyed preparations cannot always be kept safely in it because it usually becomes acid.

All preparations to be mounted in either Canada balsam or a synthetic resin must be previously totally dehydrated (sensitive objects in steps of 20%) and thoroughly soaked in its solvent—xylol, methyl benzoate or terpineol. Otherwise, instead of obtaining a beautifully clear appearance in the object through the resin, there will be ugly cloudiness. Synthetic resin which has become too thick may be thinned out with xylol again (only use absolutely clean xylol). A much quicker method than mounting in resin is to use water-soluble substances, such as glycerin or glycerin-jelly. However, resin preparations, thanks to the resin's clarity, are better than preparations made with other sealing substances.

EYES OF THE JUNE BUG

The eye can easily be removed from the head with scissors, tweezers and needles. Because this object is very thick and rich in pigmentation, the chitin must be freed from the softer parts and isolated. A simple method to destroy all soft parts is boiling the object in question in 20% to 30% potassium hydroxide solution (maceration). The potassium hydroxide loosens the tissue, and the hard coating of the eye peels off so

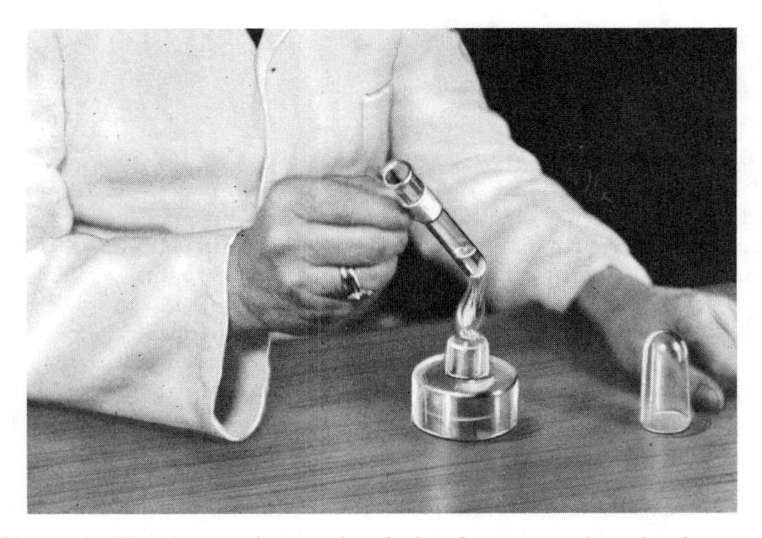

Fig. 44. Boiling in caustic potash solution for maceration of soft parts. Hold the test tube opening away from your eyes.

that it can easily be examined microscopically. Cut away the eye and put it in a test tube or small flask with a small amount of potassium hydroxide (caustic potash). Care must be used when boiling as the potassium hydroxide easily sputters out and can cause dangerous injury to your eyes. For this reason, always hold the test tube with its mouth directed away from you (Fig. 44). If the solution splashes on your hands or clothing, immediately wash with a great deal of water and neutralize with acetic acid. Violent boiling and splashing of the solution can be reduced if a splinter of wood is boiled with it.

After maceration, the object must be thoroughly washed in water (a couple of hours with frequent changing). Then place the object for at least half an hour in 95% alcohol, after which proceed as you did with the June bug's mouth.

In the preparation you will see only regular hexagons. The insect eye is indeed composed of many individual eyes (facets) and each of these hexagons belongs to one of these eyes (Fig. 45).

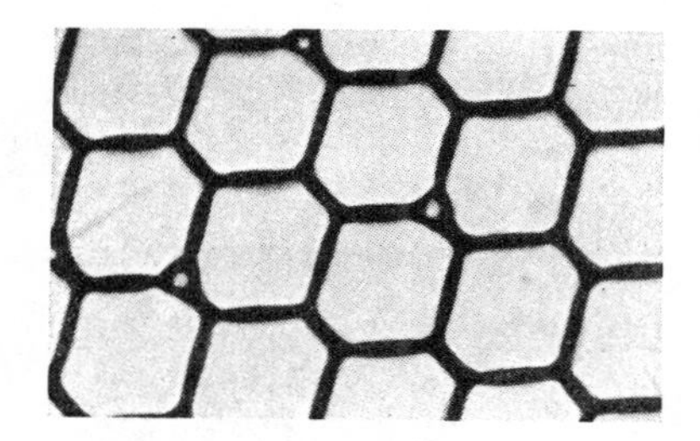

Fig. 45. Eye facets of June bug.

By means of the maceration in potassium hydroxide, you can prepare all parts of the chitin skeleton of any insect. With small insects (aphids, etc.) you can even prepare the entire insect. Boiling in potassium hydroxide is actually a "short-cut" but it disturbs many delicate fine cell characteristics. It is better to place the objects for several days, or even 1 to 2 weeks, in cold potassium hydroxide in a tightly-sealed container. The outer skin of the June bug's eye is comparatively thin and delicate. With larger objects you must naturally prolong the washing and dehydration periods considerably.

STING OF THE HONEY BEE

Cut off the terminal abdominal segment from the body of a killed bee, and separate the stinging mechanism carefully in a shallow vessel under water. The poison sac and its glands are easily separated from the hard parts if you grasp the sting with a pair of tweezers and pull it carefully away. The untreated parts should be very carefully removed from the quadrate plate on both sides of the sting, so that the sheath of the sting will not be damaged. Then place the isolated organs in spirit (2 hours) and from there into methyl benzoate (2 to 24 hours). Then mount them in resin. Hornet and wasp stings are treated in exactly the same manner.

LEGS OF THE HONEY BEE

These can be mounted in the same manner, after the bath in spirit and methyl benzoate. They are easily separated with scissors. The front leg shows the parts characteristic of insect legs: coxa, trochanter, femur, tibia and the multi-segmented foot, of which the last segment carries the claws and adhesive pad. On the upper edge of the foot you may observe the mechanism for cleaning palps which have become soiled during the search for nourishment; this consists of a cleaning notch and a spine. On the hind leg (Fig. 46), of which you mount only the lower part from the tibia on, the most interesting features are the little basket which is used for the transportation of the gathered pollen, the hollowed tibia which is trough-shaped inside, and the bristles, which appear in several rows on the first segment of the foot.

It is possible to employ the same method to prepare the interesting foot

Fig. 46. Honey bee, pollen basket.

of a housefly, the hind legs of a water beetle, a spider, and a grasshopper or the shovel-like front legs of a mole cricket.

GIZZARD OF A COCKROACH

Kill a cockroach in alcohol and cut up the underside along its entire length with scissors. On the esophagus you will see a sort of swelling and behind it a round structure the size of a small pea, the gizzard. Cut away the gizzard from the esophagus and the intestine, cut it open its full length, boil it for 10 minutes in caustic potash solution. Then wash it thoroughly in water, dehydrate in alcohol (70% and 95%) and clear in methyl benzoate. Finally spread the gizzard out on a slide and mount it in resin. (See Fig. 47.) The tips of chitin are a very pretty brown in contrast. Whenever you mount chitinous insect parts, dyeing is not necessary, because the chitin is dark enough already.

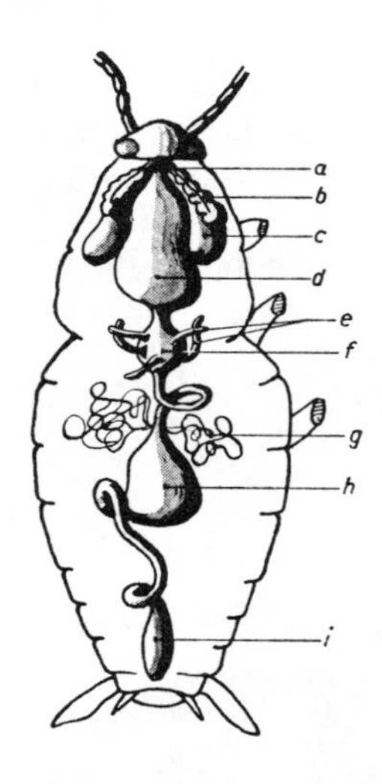

Fig. 47. Digestive organs of cockroach: (a) esophagus (b) salivary gland (c) salivary gland ducts (d) crop (e) gastric caeca (f) muscular stomach (g) Malphigian tubules (h) large intestine (i) rectum.

Usually chitin preparations are very thick and for this reason you must use a lot of mounting medium. As such preparations dry very slowly, store them horizontally for quite a time after they are completed.

FLEAS, BEDBUGS, LICE AND GNATS

Mount these in synthetic resin in their entirety, after a few hours in a 95% alcohol solution and a sufficient length of time in methyl benzoate. So as not to squash the objects with the coverslip you may have to place then in a pan 1 to 2 mm. deep, which you can easily make. For this purpose, if you do not have a glass cutter, have a glazier cut pieces of appropriately thin glass (broken slides or old photo plates) into strips 2 to 3 mm. wide and about 15 mm. long. Lay them in a square on a slide and cement them in place with synthetic resin. Fill the space inside this square with resin, mount the object in it and then cover it carefully with a coverslip. Instead of building up a square with glass strips a small metal ring may be used with a circular coverslip of the same diameter. Then heat the whole frame and allow it to cool until the preparation shows the necessary firmness. While doing this you must be careful not to heat it too rapidly; otherwise the resin will begin to boil and become filled with ugly bubbles of air. For warming, it is therefore advisable to use a warming table instead of the strong flame. This will also serve well for other purposes in microtechnique.

The construction of a warming table is simple. All that is needed is a

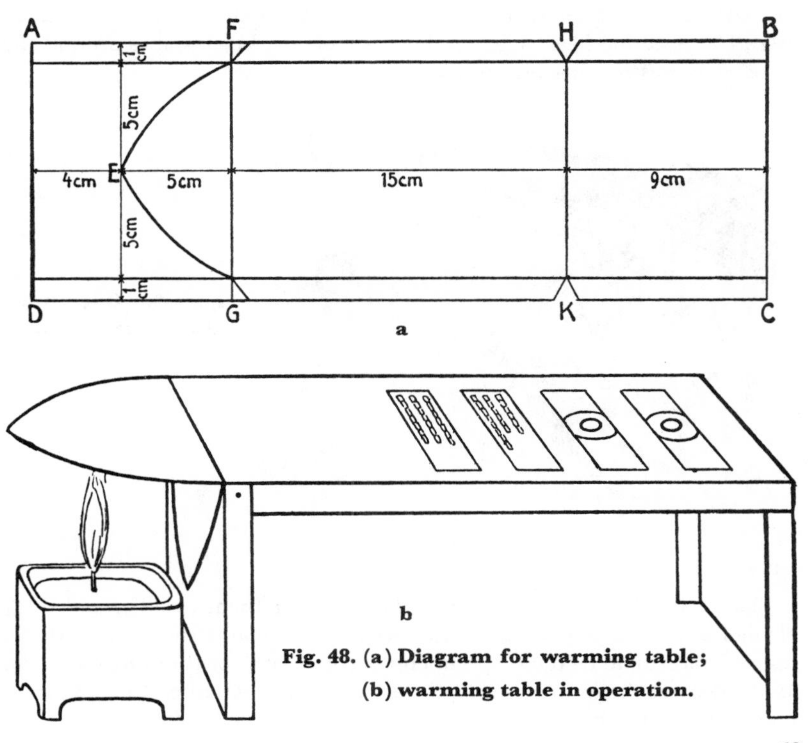

Fig. 48. (a) Diagram for warming table; (b) warming table in operation.

rectangular sheet of "tin" about 33 cm. long and 12 cm. wide, on which you draw two lines 9 cm. in from AD and BC as in Fig. 48a. Draw a border 1 cm. wide on both long sides. At H, K, F and G cut out notches. Then mark point E 5 cm. away from the midpoint of line FG, and cut a slit to form a tongue FEG. Then leaving the tongue level, bend down both ends of the table at right angles and bend over the borders. This completes the little warming table (see Fig. 48b). Under the metal tongue you can place a small alcohol or spirit lamp, which will develop a fairly steady heat. The part of the table away from the flame will always remain cooler; you can therefore use it for storing half-finished preparations, while you place the finished ones nearer the flame.

Such whole mounts are satisfying and interesting objects, which permit the study of characteristic details. Note in your flea preparation the powerful, many-spined feet; also the whole arsenal of weapons on the head, like lance-shaped cutting blades (upper jaw); the high, flattened sucking tube and two shorter cutting blades (the lower jaw) in the shape of sword blades. The tracheal openings on the body can also be plainly seen. Cat and dog fleas (Fig. 49) have a collar of black spines around the neck, which is missing on human fleas.

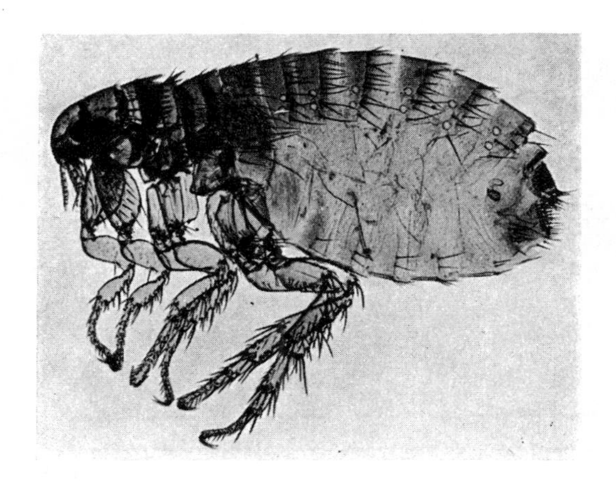

Fig. 49. Dog flea. natural size 2.81 mm.

MITES

Place mites in 70% alcohol and then transfer them to 80% and 95% alcohol and methyl benzoate, after which you mount them in synthetic resin. Unfortunately these specimens become very brittle and often the joints become bent in an unnatural manner. Therefore, the following information may help:

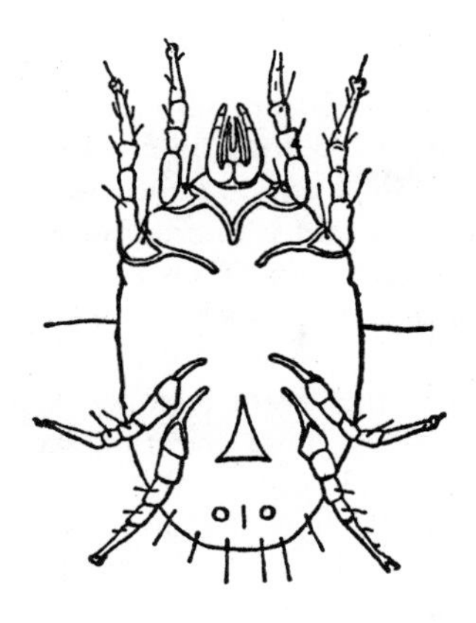

Fig. 50. Flour mite.

Immediately after you take the specimen out of the 70% alcohol bleach it by placing it in cold lactic acid (75%, as purchased). The lactic acid causes the joints to stretch. After 1 to 24 hours mount the mites in a solution called Fauré's solution which you can prepare yourself: 100 gm. distilled water, 100 gm. chloral hydrate, 40 gm. glycerin and 60 gm. gum arabic are the ingredients to be combined. Select clear pieces of gum arabic and never use the impure pulverized form. First pour the water and glycerin together, then dissolve the chloral hydrate in it, and finally the gum arabic. Gum arabic dissolves very slowly. You must wait several days and frequently shake the bottle with the solution. The thick liquid must lastly be filtered through filter paper, whereby much of the substance is lost.

Preparations in Fauré's solution must be surrounded by a varnish ring, as was used with the butterfly scales.

More easily used and easier to prepare than Fauré's solution is polyvinyl-lactophenol, a newer mounting medium.

Mix with warm distilled water enough polyvinyl alcohol to make a syrupy solution, stirring constantly. After several hours, the solution will become clear (it may be necessary to heat it again by setting it in hot water). Then add to 5 parts of the solution 2 parts of phenol and 2 parts of lactic acid. Putting a varnish ring around this solution is superfluous.

STORAGE OF PERMANENT PREPARATIONS

In new permanent slides there is always the danger that the coverslip will move out of place. The slides must therefore be stored for the first few weeks in a horizontal position, for which purpose a box for 20 slides is especially suited (Fig. 51b). For eventual storage you may use slide boxes, in which 50 or 100 can be stored upright (**Fig. 51a**). The slides must be protected from moisture, dust and light. Especially with dyed objects, long exposure to light can be very damaging. The relatively short illumination during examination is of no consequence.

Broken resin preparations can often still be saved if they are placed in xylol until the coverslips become loose. The objects may then be mounted carefully on new slides. Of course, you must work very painstakingly, because the objects are often extremely brittle.

If you make your preparations according to a determined plan, taking in the scope and content of your collection and housing it in boxes and cases, the last job is the preparation of a card index, with each slide having its separate card with room for explanatory notes. Use index cards 5 by 3 in. in size. For very large collections it is advisable to use different colored cards for each major subject category; for instance, zoological preparations (Z) red, botanical preparations (BO) green, bacterial preparations (BA) yellow, plankton preparations (PL) blue, etc. If you want to avoid this, just slip identifying cards between the categories, listing in the upper left-hand corner the name of the collection in that group; then arrange the cards alphabetically between.

Don't get "beginner's fever", working up every object into a permanent preparation, even saving the worst specimens. Carelessly prepared objects or those with which you had misfortunes find no friends. For this reason throw them away. Living examinations should take precedence; all living tissues change when they are killed and prepared. When at all possible, observe your objects alive before you make them into permanent preparations.

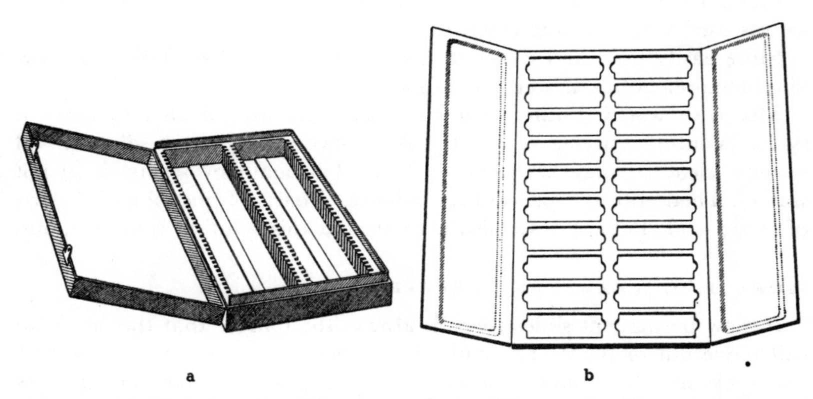

Fig. 51. (a) Slide box for 100 preparations; (b) case for 20 preparations.

5. Exploring a Drop of Water

In every standing body of water, whether large or small, lives a rich and varied world of small living creatures. So, the microscopist nature-lover need never be without material for observation. Only a few can be described here—some of a plant nature, others animal. After the different methods of observation are learned, you can always make similar observations and examinations.

DIATOMS

These are found in greatest numbers in the spring after the snow has melted, and also in the autumn, along the banks of standing bodies of water in brown, slimy masses. In the sunshine these masses rise to the surface of the water, and may be dipped up. Also, the brown coating on stones and pilings in the water consists mainly of diatoms, which can be scraped off. A useful instrument for seeking and successfully collecting diatoms is a $\times 100$ pocket microscope (Fig. 52). It is sufficient to make

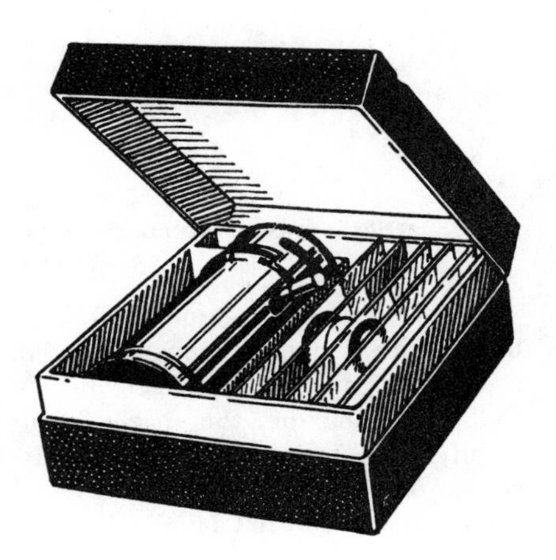

Fig. 52. A 100-diameter pocket microscope.

little tests of collected materials on the spot, thereby eliminating the danger of filling many containers with the same sort of specimens. Place the material to be examined between two thin slides of the same thickness, then place it in a slot under the lens of your pocket microscope. Then hold up the microscope to the brightest part of the sky, but not to the sun, and turn the lens back and forth until you get a sharp focus.

Examine living diatoms in a drop of water under the cover-glass; for this it is often necessary to have strong magnification and often strong cutting down of the light. The tiny, mostly boat-shaped, plant structures may be seen, moving slowly, often in a jerky manner, through the field of vision, sometimes backward and sometimes forward. They are encased in two long shells, which fit over each other like a box and its cover (Figs. 53).

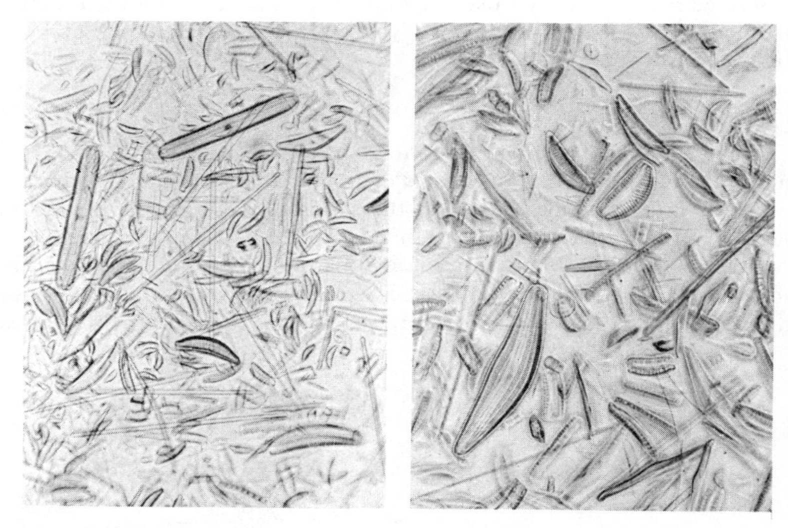

Fig. 53. Diatoms (various species); (a) magnified 80 times; (b) magnified 150 times.

In most cases the diatom-bearing material must be cleaned before it is used for a permanent preparation. For this purpose put the material into a cone- or cup-shaped beaker (Fig. 54), pour water over it and shake it well. After shaking, the coarse impurities quickly settle; then pour the cloudy water into a second beaker. Repeat this process until the settlings consist only of diatoms. The fine shells sink very slowly. The contents of the beaker should not be allowed to get warm, so as to avoid currents in

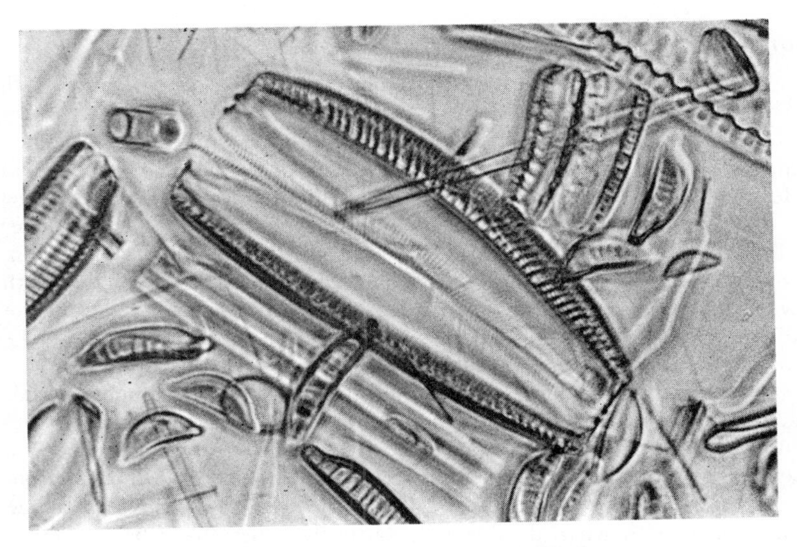

Fig. 53. (c) Diatoms, magnified 400 times.

the water. In this process, try all the settlings under the microscope, to see if there are any diatoms. Only if there are no diatoms, or very few, in the settlings need you throw them away.

Before you work any further, ask yourself if you want to examine the intriguing shell structures, or if you would find the protoplast with its chromatophores (color-bearing organisms) more interesting. In the first case the living cell content is destroyed and in the second carefully "fixed".

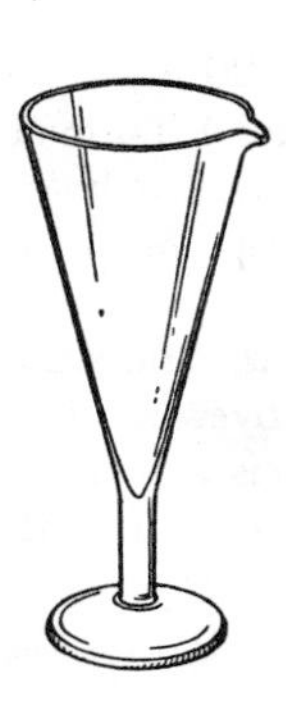

Fig. 54. (left) Conical beaker. Fig. 55. (right) Plankton net of silk gauze, usable as a throw-net or with a handle. Removable container with bayonet closure.

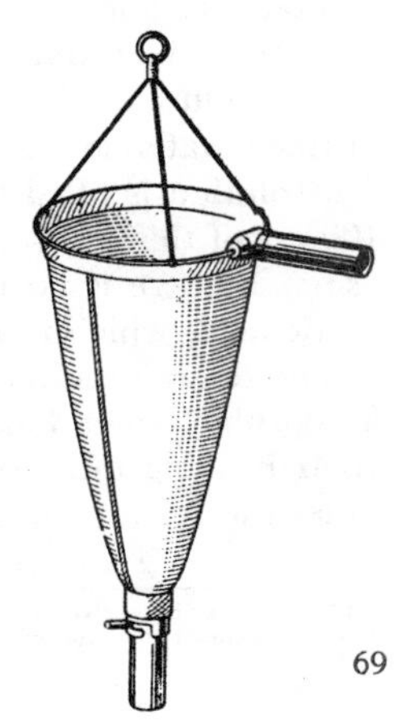

To make a shell preparation place on a coverslip a drop of water with a high concentration of diatoms; in a dust-free location allow it to dry out. Now place the coverslip—with the diatoms on top—on a mica plate and hold it over a small flame with a pair of tweezers. The heat will destroy the cell content of the diatoms. Under no circumstances allow the coverslip to be brought to a glow for it would then bend. The layer of diatoms, when heated, turns brown at first, then grey-white. From time to time, allow the coverslip to cool and check the destruction with weak magnification. If all organic substances have disappeared, and the diatoms appear clean, place the coverslip—face downwards this time—in a drop of diatom-sealing material.

For mounting such shell preparations the ordinary sealing materials are not good enough. To bring out all the delicate, many-faceted structures of the diatom shells and make them clearly visible, you need a sealing medium which has a light refractive index very different from that of the shells.[7] The simplest is an air mount which you became acquainted with when dealing with butterfly scales (page 54). Alternatively, mount in a resinous substance with a particularly high refractive index, for example, styrax.

For examination of the finest structure of details in diatom shells, an oil-immersion lens should be used (see page 135).

If you wish to examine the cell structures instead of the shells, you must not be as crude as with the shell preparations. The concentrated and washed diatom material must first be "fixed". By this is meant killing the cell contents and at the same time retaining the original form of the structures. All fixing materials are poisonous and albumen-destroying substances (but not every poison is a good fixing material!). Fix the diatoms with chrome acetic fluid, which is also useful for fixing plant cells. A 5% solution of chromic acid, which may be stored, is diluted with distilled water so that a 1% solution results. (For example, 20 cc. of the 5% solution is filled to the 100 cc. mark with distilled water.) To every 100 cc. of this 1% solution 0.25 cc. of glacial acetic acid is added before using. If there is no measuring pipette at hand, add 15 drops of glacial acetic acid, which is about 0.25 cc.

The diatom sediment in the beaker, which should hold as little water as possible, should have the chrome acetic fluid poured over it. Take enough fixing fluid for this purpose so that the fixing fluid is at least 50 times the volume of the sediment. Stir the sediment up and let it settle.

[7] On the other hand, with dyed preparations, the refractive index of the sealing material should come as close as possible to that of the tissues.

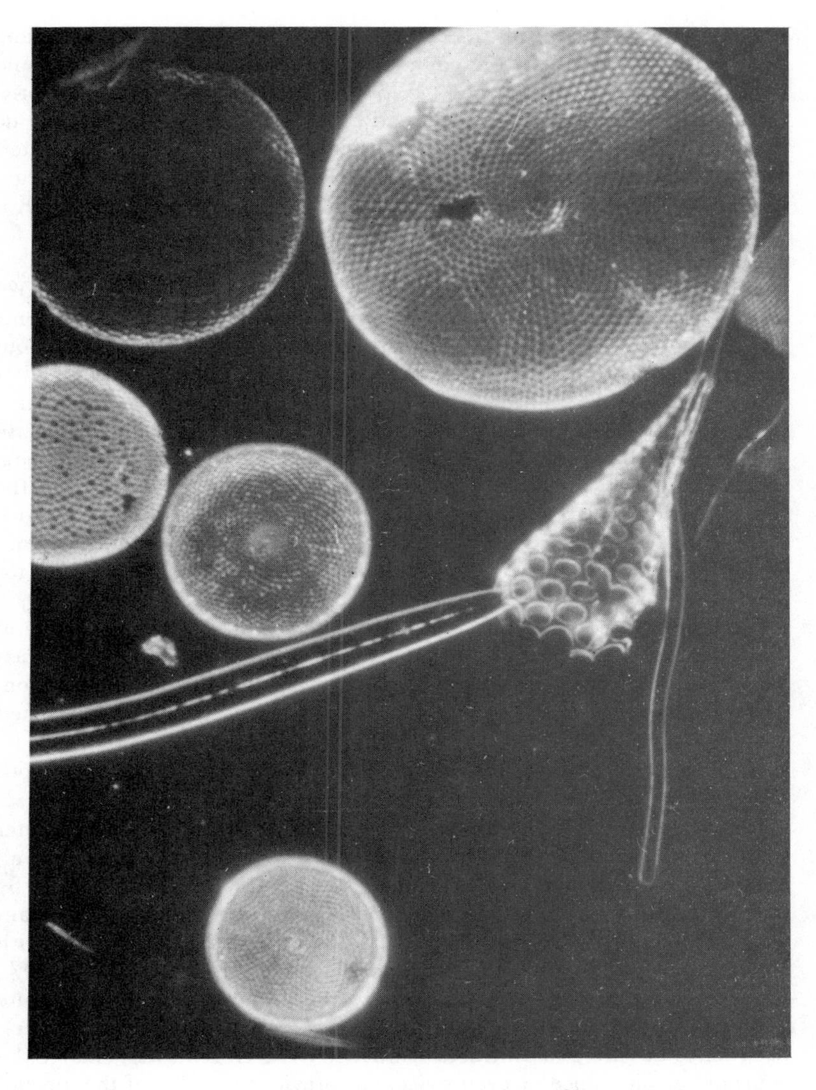

Fig. 56. Diatomaceous earth from Karand, Hungary, in a dark field.

Half an hour later pour off the chrome acetic fluid from the sediment and replace it with water. In the course of another hour the diatoms should be well washed. For this purpose, keep stirring up the sediment frequently, then let it settle, pour off the water, add fresh water, and so on. It does no harm if the fixed material remains overnight or even longer in the water.

When the water has been changed at least four times, you can begin the dyeing, and for this purpose you want to use a simple dye which is practical and does not over-dye: alizarin-viridine chrome alum, for example.

Here's how to prepare alizarin-viridine chrome alum: heat a 5% solution of chrome alum in distilled water until it boils. In the hot chrome alum solution stir in as much alizarin-viridine as can be dissolved (about ½ mg.). Allow the dye solution to cool and then filter it through filter paper. The filtered solution is now ready for use and may be kept for a long time if it is well sealed. Older dye solutions should be filtered again before use. Alizarin-viridine chrome alum is well adapted for dyeing almost all plant preparations, especially algae of all sorts. Pour the alizarin-viridine chrome alum solution over the diatom sediment and give the dye a few hours to work its way in. Because alizarin-viridine does not dye too deeply, you can let the diatoms remain in the dye solution for 24 hours without damage.

After dyeing, pour off the dye solution and wash the sediment several times to get out the excess dye. The dye solution can be filtered and then used again. You must be very careful when pouring off the dye solution. Alizarin-viridine is so dark that the sediment can easily be overlooked and poured out with it.[8]

Now the dehydration follows. This can be accomplished by various methods:

(*a*) *Progressive dehydration with alcohol solutions.* Pour off the rinsing water from the bottom sediment and replace it with 35% alcohol. So that all diatoms are surrounded with alcohol, stir up the sediment (either by shaking or by stirring with a glass rod). When the material has again settled, pour off the 35% alcohol after 3 to 5 minutes and replace it with 70% alcohol. In the same manner place the material successively in 80%, 90% and 95% alcohol. Finally, replace the 95% alcohol with absolute isopropyl alcohol, which should work in for at least 10 minutes (or longer). In order to be sure that all traces of water have been removed, change the isopropyl alcohol once or even twice. All previous remains of the alcohol with a water content are wiped from the glass rod or dissolved in the

[8] If alizarin-viridine is not available, the diatoms may also be dyed with Delafield's hematoxylin. The method is described on page 103.

first isopropyl alcohol stage. The isopropyl alcohol is followed by xylol, which should work in for at least 10 minutes and also be changed once. Now place on a slide the sediment which has been soaked with xylol, and mix it with a drop of synthetic resin and cover it with a coverslip. It is important that the material does not become dried at any time during these processes.

To achieve the same ends you might save time by placing your diatoms immediately in 100% alcohol for dehydration. However, by taking this sudden step from water to 100% alcohol such delicate cells as diatoms would shrink greatly. For this reason it is much better to perform the dehydration in gradual stages.

By using the following procedures with glycerin or with methyl glycol you require a fewer number of stages.

(*b*) *Dehydration with glycerin.* The glycerin procedure takes longer, but presents considerably fewer working stages. Dilute glycerin with distilled water at a ratio of 1:6 to 1:10 and pour the diluted glycerin into a watch-glass. The dyed and washed sediment of diatoms is taken from the beaker and placed in the glycerin-water solution in the watch-glass.

Now wait until the water evaporates and the glycerin becomes thicker. Depending on the temperature, this can take several days. For this purpose place the watch-glass uncovered in a dust-free location or in a box. You can only continue your task when the fluid in the watch-glass is again about as thick as the pure glycerin. Because in this process the fluid mass becomes considerably reduced, you immediately scrape together the sediment in the middle of the watch-glass; otherwise it could happen that the main mass of the material would finally dry out and stick to the edge of the watch-glass.

Take the diatoms directly from the thickened glycerin and place them in absolute isopropyl alcohol, which must be changed two or three times to get out the glycerin. For this purpose again work with a beaker. As already described above, follow this with one or two xylol stages and finally mount it in resin.

(*c*) *Dehydration with methyl glycol.* Methyl glycol dehydrates so delicately, that you are able to do your task in very few stages. After dyeing, replace the water over the diatom sediment with a 50% methyl glycol solution (methyl glycol and distilled water in equal parts), and allow this to work in for about 5 minutes. Then pure methyl glycol follows (15 to 30 minutes, changed once or twice), xylol (change once), then the resin seal.

It should also be mentioned that with all methods—with the exception of clean shell preparations—this preparatory work does not pay unless there is a rich stock of material at hand. At least at the beginning you must figure that about 75% of the diatoms you start with will be lost. (The work will be made easier and better with the use of a simple hand-centrifuge. When working with a centrifuge, both centrifuge containers must be of equal weight.)

Fossil diatoms exist in diatomaceous earth. You can purchase it from firms which use it to insulate heating pipes. The material is cleaned as described above and mounted as a shell preparation.

SPIROGYRA

Spirogyra are found during the warm months in standing water in the form of green, slimy pads. Upon examining such a growth of algae a little more closely, you will see that it consists of many delicate little threads, which are composed of separate closely-lined-up cells. In each cell you find a spirally-wound green band, the chromatophore (color body). This band represents the well-known chloroplasts of the higher plants. Because of it the plant is known as the "spiral alga". (Fig. 57.)

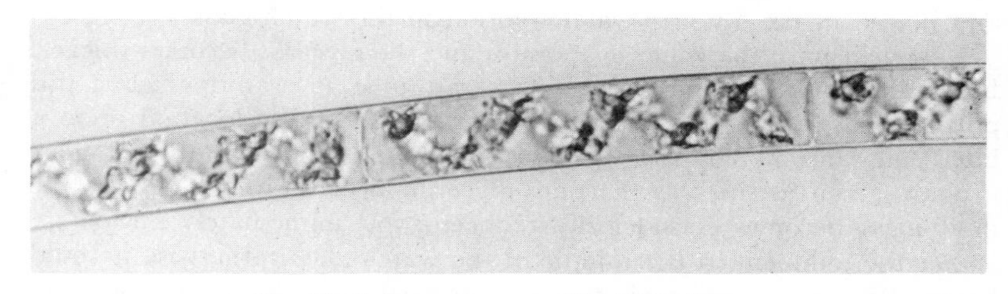

Fig. 57. Spirogyra, magnified 450 times.

Naturally you will also want to see other algae and you will find many forms here.

If you have an aquarium that contains algae you need only place some coverslips in the water and, depending upon the lighting, leave them there from 3 to 8 days. After this time carefully remove them from the water with tweezers, then allow the excess water to drip off and lay them on clean slides with the algae side down. Wipe the upper side of the coverslips dry with a bit of cotton or filter paper so that the objective lens will not be smudged.

You will find that many and various algae have colonized on the coverslips; you can almost always find green algae, diatoms and blue algae.

If you wish to make a particularly well-populated coverslip into a permanent preparation, take steps similar to those used in diatom preparations. The following preparation instructions will apply not only to algae-covered coverslips but also to thread algae, for the preparation of which you pluck out a few threads.

First fix the preparation in a sealed container of chrome acetic fluid for 15 minutes to 2 hours (see page 70). It is important here also to see that the volume of fixing solution exceeds that of the objects by at least 50 times. Wash off the chrome acetic fluid thoroughly with water and place the material in a watch-glass with highly diluted glycerin (one part of glycerin to 10 parts of distilled water). If the glycerin has become thickened, use fresh glycerin. If you were to place the algae immediately in pure glycerin, there would be ugly shrinkages. Mounting in glycerin will be explained more fully later.

The preparations will become more attractive if you dye them before enclosing them. The fixed and thoroughly washed algae are dyed for a period of a few hours to one day with alizarin-viridine chrome alum (see page 72), then washed thoroughly and transferred to the diluted glycerin. Mounting in glycerin jelly or glycerin is possible with this dye. (Most other dyes do not last in glycerin jelly.)

MOUNTING IN GLYCERIN OR GLYCERIN JELLY

Glycerin and glycerin jelly were formerly very popular as mounting mediums; they are still used extensively today in spite of being crowded out by improved resin-mounting methods.

Glycerin and glycerin jelly have the advantage that they are water-soluble; objects therefore need not be very carefully dehydrated—if at all.

Disadvantages of glycerin and glycerin jelly: the preparations require a very careful, absolutely tight seal with a varnish or a resin. Only a few dyes can be kept in glycerin and glycerin jelly.

As a rule glycerin jelly is preferable to glycerin, because glycerin jelly stiffens when it cools. The following instructions for glycerin jelly also hold good for glycerin, but for glycerin the heating is eliminated.

Glycerin jelly is best purchased ready for use as it does not pay to make it up. Take from the bottle with a spatula, or with a small spoon, as much as you can predict will be necessary for immediate use. Place the pieces in a watch-glass or, better yet, in a small test tube. In a bath of moderately hot water the glycerin jelly will become fluid. When it cools again to body temperature you can work with it. The reserve supply should not be made fluid, because repeated heating and cooling makes it brittle.

For the mount you need, besides the glycerin jelly, small coverslips and larger coverslips as well as thin-flowing synthetic resin (it may be advisable to thin the resin used here with xylol). On a small coverslip put a drop of glycerin jelly—just enough to fill (later) the space underneath the coverslip, never more. In the drop place the object after taking it from the thickened glycerin (see page 73). Now grasp the small coverslip with your tweezers, turn it over quickly so that the glycerin drop with the object hangs underneath and lay the small coverslip exactly in the center of the larger coverslip which has been prepared. Now the object lies—mounted in glycerin jelly—between the small and large coverslips. On a slide place a heavy coating of resin on an area which roughly corresponds to the area of the larger coverslip. Grasp the larger coverslip with tweezers and turn it over quickly; the small coverslip which is lying on it is now hanging underneath. Lay both slips—the small one underneath—flat in the resin. The resin immediately flows around the small coverslip and seals off the glycerin jelly completely (Fig. 58). If the mounting resin does not fill the space under the large coverslip completely, flow more resin in from the edge. It is important that the small coverslip is not moved out of place when laid in the resin; otherwise there will be cloudy areas at the edge, where the glycerin jelly is forced out from under the small coverslip.

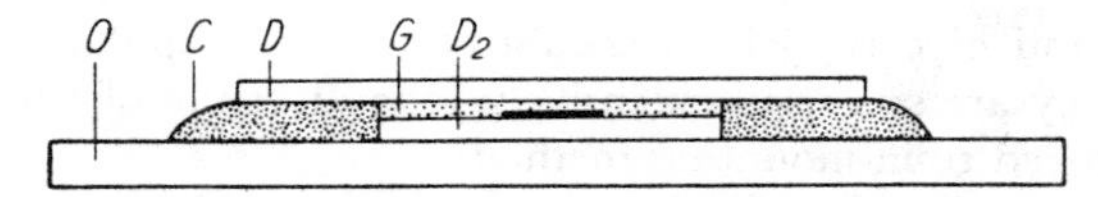

Fig. 58. Glycerin mount. (D) Large cover-glass. (G) Glycerin drop with object. (D2) Small cover-glass. (O) Slide. (C) Resin.

For very small objects you can use in place of the small coverslip a fragment of a broken coverslip and instead of a large coverslip one of normal size.

The described method may seem inconvenient. With a little practice, however, it is easier to use than the older and seemingly simple procedure with a varnish-ring seal.

If you seal with a varnish ring, place the drop of glycerin jelly right on the slide, and cover it with the coverslip. Here you must also be careful that no trace of the glycerin jelly is pushed out from under the coverslip. After the glycerin jelly cools and stiffens, add a ring of varnish. Apply the ring so that it covers the edge of the coverslip, and connects the coverslip and slide without any holes, thus sealing in the glycerin jelly completely. Once the varnish has dried apply a second coat, then a third, and if necessary, a fourth.

Drawing neat varnish rings freehand is very difficult. It is therefore better to use a special turntable for the purpose. Clamp the preparation on the turntable and turn it slowly while applying the brush in the spot where you intend the ring to appear. The use of such a turntable naturally requires round coverslips.

Some microscopists also put these rings around synthetic resin preparations, because they feel that they give the preparations a more attractive appearance. A varnish ring is, however, not necessary with a resin preparation.

Glycerin jelly preparations (not glycerin preparations!) may be kept for several months without being specially sealed.

WATER FLEAS

These occur in every pond in large quantities. Catch some in a little sieve, such as aquarists use. It is simpler still to purchase a few water fleas from an aquarium supply dealer. You can easily learn from water fleas how to preserve plankton in permanent preparations.

Water fleas are very active and move out of the visual field all too easily. It is helpful to examine them in the following manner: Make wax feet for the coverslip from adhesive wax (Fig. 59) so that by pressing down on the coverslip you can pin down the organisms. If you do not want to use this method, try a little quince mucilage added to a drop of fluid to slow down the fleas' movement. You can make this yourself by pouring some water over a few crushed quince seeds in a test tube and letting it stand a few days.

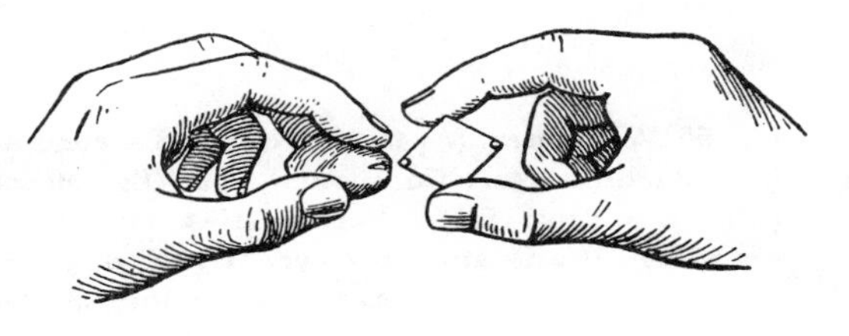

Fig. 59. Making wax feet from block of wax.

Still simpler is the provision of an impeding substance with which the motion of water-life (including *Paramecium*, etc.) can be slowed down. Such a substance is methyl cellulose which may be purchased from any biological supply house. Take 3 gm. of methyl cellulose and pour over it 95 cc. of warm water, stirring it well and letting it stand overnight, after shaking it frequently. The result is a tough slime, which is stirred into the drop of water which contains the aquatic life. Motion is slowed down, depending upon how much of the impeding mixture is added.

Very often all that is necessary is to draw off some water from the edge

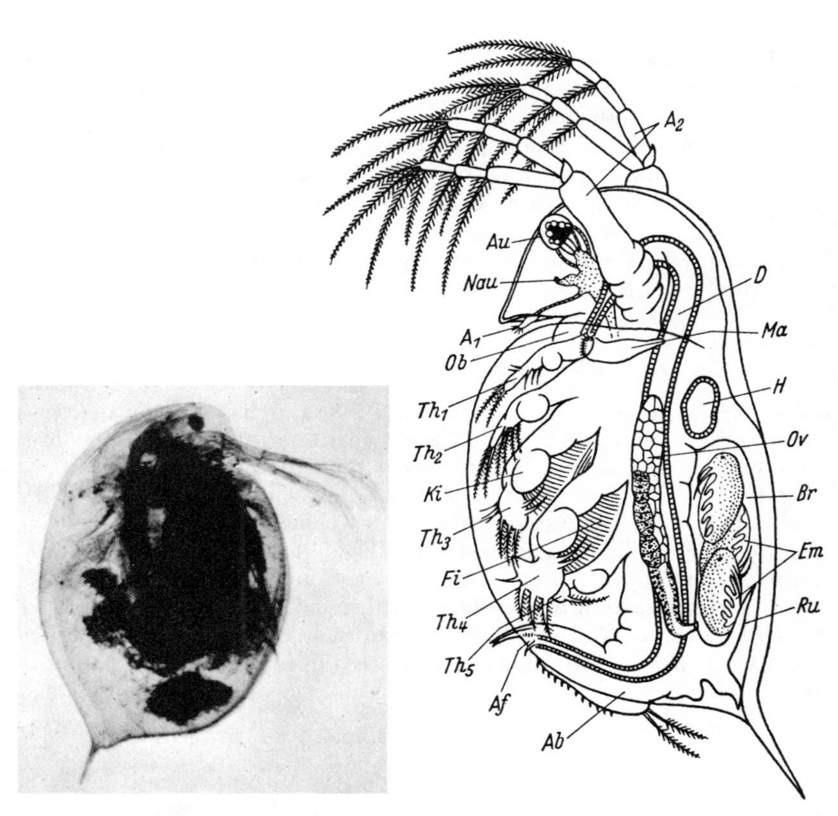

Fig. 60. Water flea. (A_1 to A_2) First and second antennae. (Ab) Abdomen. (Af) Rectum. (Au) Compound eye. (Br) Brood pouch. (D) Intestine. (Em) Embryos. (Fi) Filter bristles. (H) Heart. (Ki) Gill sacs. (Ma) Mandible. (Nau) Auxiliary eye. (Ov) Ovary. (Ru) Dorsal continuation. (Th_1 to Th_5) Swimming feet.

of the cover-glass in order to arrest all movement. Even with small one-celled organisms this procedure can be used.

To make permanent preparations, place several water fleas in a solution consisting of one part of 40% formaldehyde (formalin) and four parts of tap-water (formalin 1:4). The organisms are fixed in this. Because of their small size, they will probably be fixed after a half-hour, but a longer immersion does no harm. Then wash the water fleas well in frequently changed tap-water (at least a half-hour) and then transfer them to distilled water. To dye them you can use hemalum or borax carmine. Hemalum dyes very rapidly, but dyeing with borax carmine gives more attractive results.

To dye with hemalum put the water fleas in a receptacle with Mayer's acid hemalum. The length of time for dyeing depends upon the size of the objects and the age of the dyeing mixture. Usually you can get away with a 10 to 15 minute immersion. Then there is a washing period in frequently changed tap-water for 10 to 20 minutes, during which the color changes from reddish at first to a deep blue. If the color is still too pale, transfer the water fleas to distilled water and then back to the hemalum. If the color is deeper than you expected, a part of the dye may be drawn out with an acetic acid-alcohol solution. (To make this solution: to 100 cc. of 70 % alcohol add 0.5 cc. of pure acetic acid. Do not use industrial spirit.) Allow the organisms to remain in this solution for about a minute, by which time the color will have returned to red; then place them in ammonia water (a few drops of ammonia to 50 cc. of water). Finally wash them thoroughly in tap water. This prolonged process is, however, usually unnecessary when dyeing with hemalum.

The dyed and water-flooded objects should then be dehydrated by the alcohol stages or with methyl glycol without stages. When using the alcohol stages for dehydration, begin with 35% alcohol and end with 70, 80, 90 and 95% alcohol. At every stage allow the water fleas to remain in it for at least 10 minutes. Follow this with two changes of methyl benzoate, in which the objects remain long enough until they sink to the bottom and become pale (at least 15 minutes) and finally mount them in synthetic resin. In order not to crush the water fleas, slip in a few slivers of broken cover-glass.

Very much faster than alcohol stages and just as surely effective is dehydration with methyl glycol. Transfer the water fleas from the water to methyl glycol for 15 to 20 minutes, and change it once. From here transfer them directly to methyl benzoate. Further steps are as previously described.

To dye with borax carmine, place the fixed and washed water fleas first in 35% alcohol and next in the borax carmine solution. Here let them remain for 12 to 24 hours (cover the dye solution tightly) and then put them in a salicylic acid-alcohol solution (see above). This solution "differentiates", that is, it draws some of the dye out again. Differentiation time varies depending upon the nature and size of the object. You control the amount of differentiation by the length of immersion. Once the organs are sharply outlined when placed under the microscope in 70% alcohol, stop the differentiation. In other cases the objects should be placed in the salicylic acid-alcohol once more. Several hours may be necessary for the differentiation process, and even days for larger objects. Interrupt the process by washing for several hours in 70% alcohol, changing it three or four times. (All acid traces must be thoroughly removed.) Finally pass the objects through 80, 90 and 95% alcohol, then methyl benzoate, and then mount in synthetic resin.

MICROAQUARIA AND HAY INFUSIONS

The microaquarium can be an ideal source of objects for observation, and can give you a deep insight into the laws of life and its processes. As a container you can use preserve jars or accumulator tanks or the like. Take the water to fill these containers from the same sources from which your water-plants come, namely a pond or slowly-flowing stream. For plants you can use pondweed, waterweed, hornwort or others. Cover the bottom with clean, washed sand about 1½ in. high. To the aquarium add the water which flows out of squeezed-out mats of algae and other water plants, surplus plankton, scrapings from old wood that has been lying in the water, the slime coating of stones in the water, etc. There should not be an overabundance of organisms in the containers. Avoid overheating and provide with diffused daylight. Place a glass plate on top, using a few pieces of wire at the edges, to allow the entrance of air. Suspend small pieces of raw meat deep in the containers with thread and take them out again after 24 hours. Use no snails. Evaporated water should be replaced with more from the same source. Such aquaria hold up for years and are a constant source of material for examination and research.

A hay infusion is nothing more than a specialized microaquarium. Its main purpose is to supply Infusoria, which are so called because they are easiest to find in such infusions.

Stuff into a preserve jar a handful of hay or straw and fill it with pond or aquarium water. Allow this container to stand open for several days. The material will begin to decay, and a film will form on the water. In this film as well as in the water only bacteria will be found at first, accompaning the decay. Soon you will also find some single-celled organisms (Protozoa[9]), at first small ones and then larger Infusoria, and finally Amoebae. For Vorticellas, a better medium is an infusion of fresh parsley, grass or cauliflower, and for the beautiful Stylonichias the best medium is an infusion of locust blossoms.

It is best to keep water fleas away from infusions, because the cultures

[9] One-celled animals are also quite easy to find in nature. They occur in ditches, in manure piles and by the million in any more or less dirty waters.

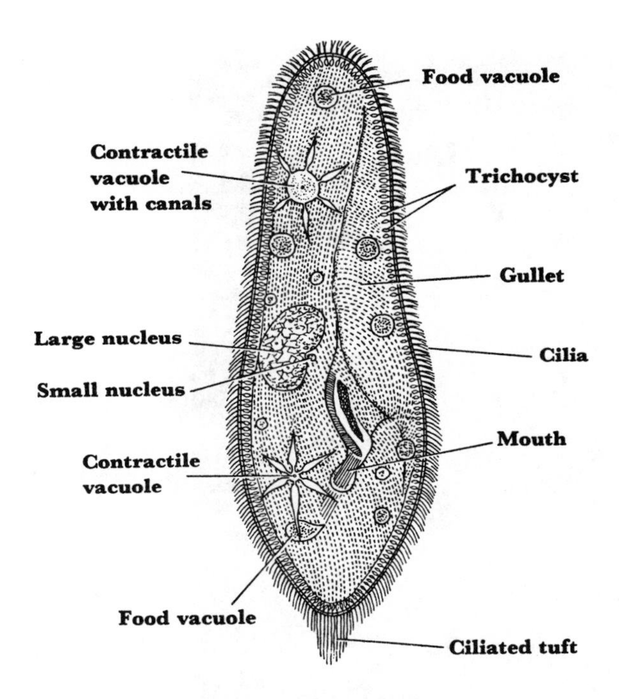

Fig. 61. Paramecium.

must be protected from these predators. Place the culture containers in a not too bright location and take tests daily. Record the occurrence of the different organisms by drawings or in permanent preparations.

PARAMECIUM

The *Paramecium* is an animal classified with the Infusoria because of its structure (Figs. 61, 62). It propels itself by means of cilia. When the puddles in which it lives dry out, it falls into a kind of sleep, surrounds itself with a hard covering (cyst) and in this dry state is carried away by the wind. When conditions are again favorable it awakes from its "encysted" state and "comes back to life". This life cycle is shared by all other Infusoria.

You will be certain of getting hordes of paramecia if you add a little meat extract to a hay infusion and then "inoculate" with a little pond water or aquarium water.

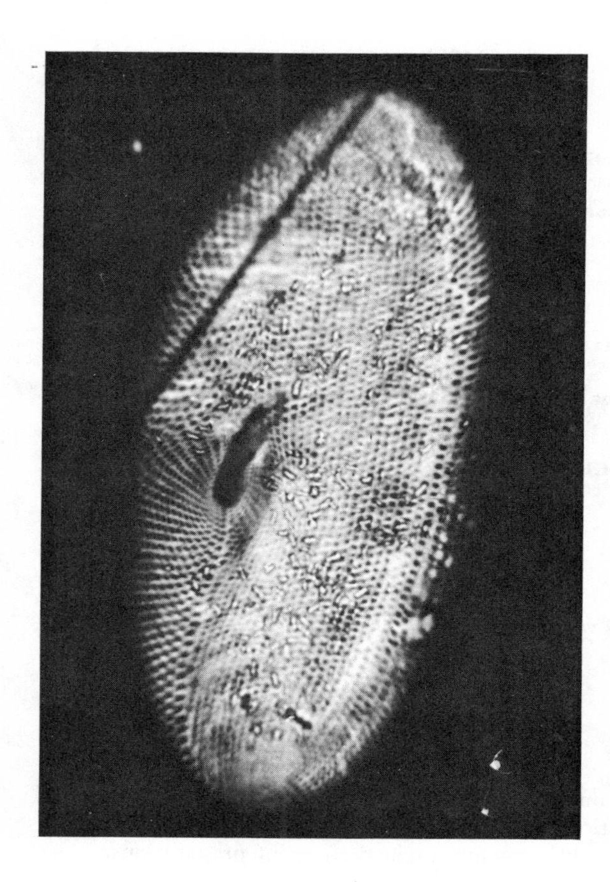

Fig. 62. Paramecium from a smear preparation.

From this or another culture place a drop with a great many animals on a slide. Observe the paramecia with medium magnification, as they gather around the edge of the coverslip and around air bubbles because they require a great deal of oxygen.

To another drop of paramecia, add a few grains of carmine and cover with the coverslip. The carmine is stirred up by the animals, and as a result the feeding vacuoles are filled with carmine grains.

The making of permanent preparations of Infusoria is very difficult, unless you make use of a very simple method. This method can give exceptionally good results; above all, the outer structure of the cell bodies will be shown with a clarity that could not otherwise be achieved. Unfortunately the method is not always successful; sometimes it is necessary to make up five or ten preparations until one is really satisfactory.

To carry out this "smear" method, place a small drop with as many Infusoria as possible[10] on a slide. Alongside of it place a drop of prepared opal blue phloxin rhodamin solution. Quickly mix together both drops with the help of a glass rod or a so-called coverslip spatula (Fig. 63).

Fig. 63. Slide spatula.

Then spread out the well-mixed fluid on the slide in a not too thin layer. The dye penetrates the finest recesses, so that the surface sculptures, cilia, etc., appear with surprising clarity. It is important for the smear to dry quickly. For this purpose put a hair dryer or fan to work, using the cold air stream to accelerate solidification. (Do not use a hot stream of air.) Cover the fully-dried smears with a coverslip and mount in synthetic resin. This is one of the few instances where a preparation may be dehydrated simply by drying it.

ROTIFERS (WHEEL ANIMALCULES)

These animals occur in almost all waters. They are often confused with Infusoria by the beginner, but are multicelled and closely allied to the worms. They are called "wheel animalcules" because of an outer and inner circle of hairs at the head end, which is in constant motion and gives the impression of a wheel; its purpose is to fan particles of food into the creature's mouth. The animals are often provided with attachment organs, which resemble hornlike protuberances, or long spines which stand out from the surface.

You can get a rich source of material for observation if you squeeze out some plants from an aquarium, or damp moss pads from the woods, trees or roofs into a watch-glass, pass the water which flows off through a plankton net and wash the clinging organisms from the silk gauze into a

---

10 In order to get the greatest number of Infusoria, confine yourself to the places where they get the greatest amount of nourishment: accumulations of decayed substances. Therefore take with the forceps a small bacterial accumulation (the floating film) which you can see with the naked eye as whitish particles on the surface of the culture water, and extract from this a drop of water on your slide. Many Infusoria can also be obtained by removing a bit of hay or straw from the water with a pair of forceps and drawing off a drop of water with another pair of forceps. If things still do not come up to expectations, then enrich the paramecium content. To do this pour some of the contents of your culture jar through a piece of cheese cloth into a beaker (Fig. 55). Because the filtered fluid is poor in nutritional substances—all coarser plant fragments and accumulated bacteria are sifted out—the organisms in this will remain alive for a few days only. Fill a test tube with this fluid and place it upright in a warm spot. After a short time the organisms, which require a great deal of oxygen, all have gathered at the surface. You have thus a convenient method for obtaining a very Infusoria-rich fluid.

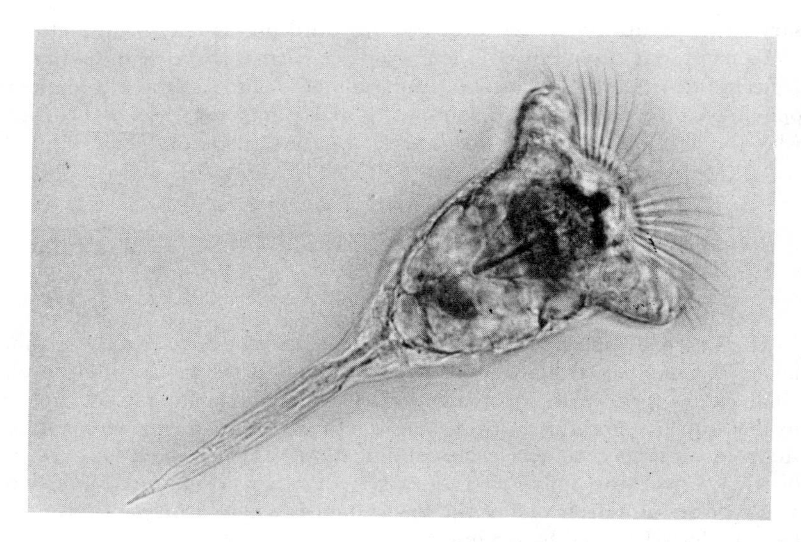

Fig. 64a. Rotifer enlarged 45 times.

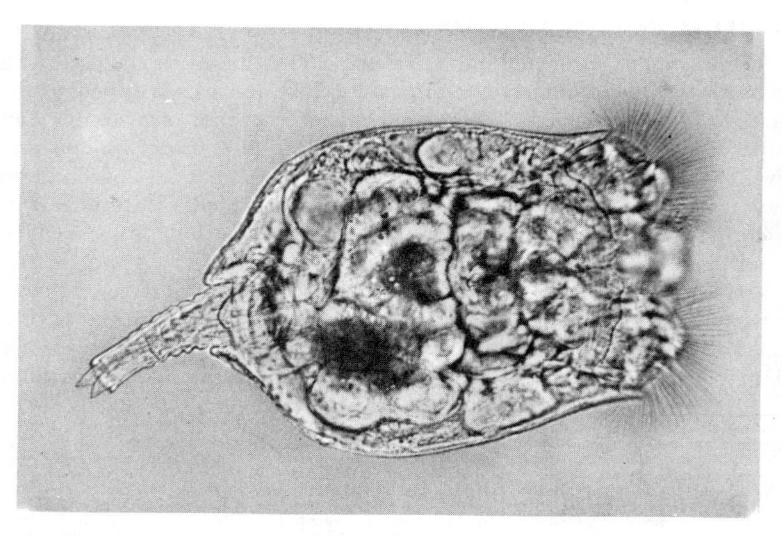

Fig. 64b. Rotifer.

small quantity of water. From this you eventually take a small drop and place it on a slide.

To see the unusual wheel organ and the digestive organs, set up for preparation some quince mucilage or methyl cellulose (see page 77). Naturally you can provide the coverslip with wax feet and then by careful pressure try to check the rapid motions of the animals. It is of course obvious that you first observe the living organisms. With preserved specimens it is usually possible to see only the armored forms well.

To make a permanent preparation you must first stun the organisms, so that when they are fixed they will not contract and become altered beyond recognition. For this purpose, use carbon dioxide in the following way.

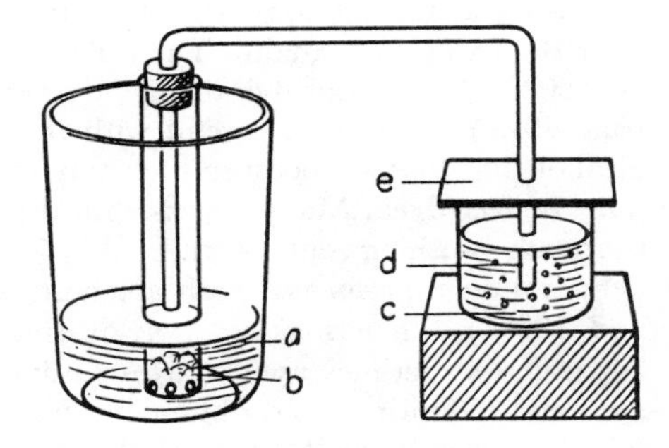

Fig. 65. Apparatus for anaesthetizing rotifers with carbon dioxide.

In a test tube with a perforated bottom place a few pieces of calcium or marble. Seal with a cork into which a hole has been drilled and into which you insert a bent glass tube (Fig. 65). Stand the test tube in a beaker which contains a dilute hydrochloric or 20% salicylic acid (*a*). When the acid comes in contact with the calcium pieces, carbon dioxide is evolved. Through the glass tube (*d*), the gaseous carbon dioxide passes into the vessel which contains the water with the rotifers (*c*). This vessel is covered with a sheet of cardboard (*e*). The carbon dioxide stuns the rotifers, a process which takes 15 to 40 minutes.

To the stunned rotifers add enough 40% formalin to make the strength of the entire solution about 4% or 5%, and let it work in for half an hour. The rotifers are of course microscopic, and for this reason you cannot simply grasp them with tweezers and transfer them from one fluid to another. Therefore, for further processing you must either apply the method used with the conical beaker for the diatoms, or use a centrifuge.

A very valuable apparatus in all preparations of the smallest objects is a simple hand centrifuge. In principle the centrifuge works by centrifugal force to achieve the settling operation as in the pointed beaker. However, the material settles much more quickly. This has two pockets in which two centrifuge glasses, joined at the bottom, are placed. These two glass receptacles must be of equal weight when filled. If only one is filled with a liquid, then the other must carry an equal weight of tap-water. Unequal distribution damages the centrifuge and makes the "throwing out" of the little particles more difficult. Choose for plankton, algae and the like, a medium rate of revolution for the centrifuge, so that they will not be damaged by the action. When the material for observation has settled in the bottom of the centrifuge glasses, pour off the surplus liquid and replace it with new fluid. Add only a few drops and shake it up so that the material will be thoroughly mixed. Once the particles are well stirred up, pour in the rest of the fluid.

Wash the rotifers for half an hour in several changes of tap-water after the fixing in formalin. Then place them for 5 minutes each in 35% and 50% alcohol and dye them for several hours in borax carmine (see page 79). You must be careful with the differentiation in salicylic acid-alcohol that follows, because such tiny organisms can sometimes be very quickly over-dyed. Many checks with the microscope are necessary. After thorough washing out of the salicylic acid-alcohol in 70% alcohol, dehydration is performed with ascending strengths of alcohol (80, 90, 95 and 100%). Then soak with xylol and mount in resin. With rotifers euparal is sometimes used as a mounting medium. Euparal is a type of synthetic mounting resin obtainable from biological supply houses. It is soluble, however, in 100% alcohol, and for this reason you can eliminate the xylol stage and go directly from absolute alcohol into euparal.

EUGLENA

This is a flagellate organism found in a pond or in a long-standing plant infusion. The organism is provided with an eye-spot, which is recognized by its red color (Fig. 66). Since the flagella are usually recognizable only after the animals have been killed, add a drop of potassium iodide solution (or, lacking this, tincture of iodine) under the coverslip. Euglena is green and has many other characteristics of plants; it is therefore grouped with the algae. Permanent preparations are completed as with spirogyra (see page 74).

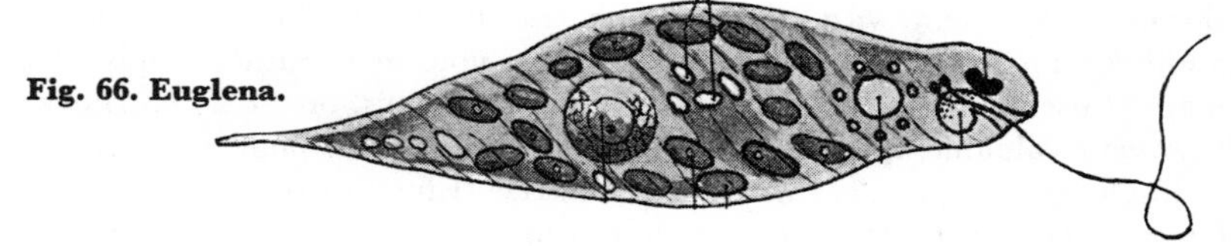

Fig. 66. Euglena.

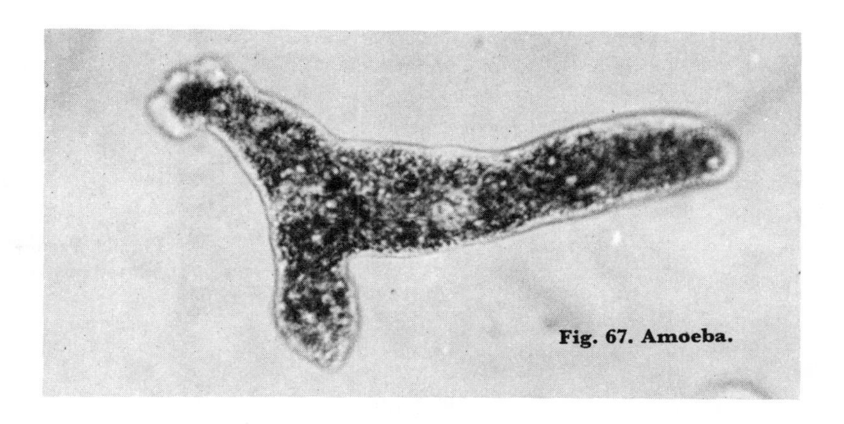

Fig. 67. Amoeba.

AMOEBAE

Naked amoebae are found on the bottom of ponds where many decaying plants are present. You will find smaller numbers in a hay infusion, if you study the developing life therein. The amoebae belong to the simplest organisms: naked, nuclei-containing drops of protoplasm (Fig. 67).

Observe the pseudopodia (false feet) which serve to propel it. In addition to the cell nucleus, a pulsing vacuole may also be seen.

Shell-bearing amoebae (Fig. 68) are numerous in black, evil-smelling mud from pond bottoms. Above all you will find them on the bottom of all waters where there are stones and shells in addition to sufficient nourishment.

An important instrument in this preparation is a pipette which has been drawn out to a fine point, for transferring matter from one test tube to another. The transfer must be successfully performed under the microscope.

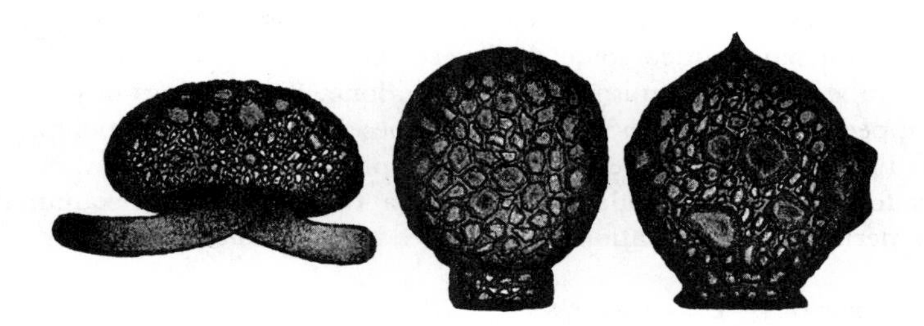

Fig. 68. Shells of shell-bearing amoebae.

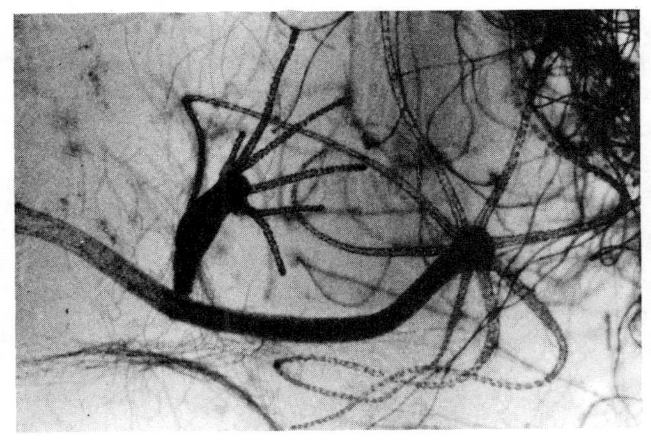

Fig. 69. Fresh-water polyp.

To make permanent preparations, take a test sample which is rich in organisms, put it into a glass dish and fix it for 5 to 10 minutes in a mixture of 9 parts of concentrated aqueous picric acid solution and one part of acetic acid. (The picric and acetic acids should be thoroughly mixed before using.)

At the opening of a pipette, tie a piece of silk gauze, of the kind used for plankton nets, and draw out the fixing fluid, immediately replacing it with 70% alcohol. To get out all of the fixing fluid change the 70% alcohol several times in this manner. Dye with borax carmine (see page 79). Perform dehydration by the rising alcohol stages; mount in euparal (see page 86).

FRESH-WATER POLYP (*Hydra*)

If you examine with a pocket magnifying glass a microaquarium to which you have added some duckweed you will—if you are lucky—see some tube-shaped organisms about 1 cm. long, the openings of which are equipped with a number of tentacles. These are fresh-water polyps (Fig. 69). Pick up one of these polyps with a pipette in a drop of water, put wax feet on the cover-glass and make a thorough living examination. The permanent preparation is completed as with water fleas.

PREPARING DYE SOLUTIONS

You can purchase many dye solutions ready for use; some solutions are so difficult to prepare that even the experienced microscopist will buy them. Usually it is cheaper if you make up your own dye solutions from dye substances. To do this you must have a slight acquaintance with the fundamentals of microscopic dyeing.

Microscopy uses both direct dyeing substances and those requiring mordants.

Direct dyeing substances dye without previous preparation or simultaneous preparation with a tanning substance. They are classified as acid dyes and alkaline dyes. Alkaline dyes (such as methylene blue, safranin or gentian violet) dye cell nuclei and other "basophilic" (that is, compatible with alkaline) cell structures. Acid dyes (such as eosin, orange G., light green, aniline blue) dye the "acidophilic" (acid compatible) cell structures. Because the alkaline dyes have a greater action on the nuclei, and the acid dyes have a greater dyeing action on the cytoplasmic structures, one speaks of nucleic dyes and cytoplasmic dyes. Most solutions with direct-acting dyes can be prepared in the following manner:

In 70% alcohol (use pure alcohol—not industrial spirit—isopropyl alcohol may be used in a pinch), dissolve enough dye powder so that in spite of repeated shaking there will be a sediment of undissolved dye on the bottom. This saturated solution is the "stock solution", which can usually be stored for a long time. For use, dilute a part of the clear stock solution with distilled water or 70% alcohol at a ratio of 1:10 or still stronger. Remember this rule of thumb: concentrated solutions dye quickly, but often unevenly; highly diluted solutions dye slowly, but in most cases clearly and cleanly. The preference of the microscopist determines how strong he makes his solutions. The correct dyeing time is easily arrived at by checking through the microscope. If one has dyed too much, it is almost always possible to remove a good part of the dye with distilled water or alcohol.

Mordant dyes do not dye until they have been previously prepared with another substance, the mordant. An alum solution usually serves as a mordant, forming a durable dye enamel with the dye. Mordants and dyes may be used separately or—as happens more frequently—in a single compounded solution. In dyeing, the mordant dyes act here like alkaline dyes. Mordant dyes used particularly often are the hematoxylins, the carmine solutions and the synthetic mordant dyes (alizarin cyanine, alizarin viridin, etc.). In the preparation of solutions with mordant dyestuffs the directions (weight, volume, etc.) must be followed exactly. If you do not have a sensitive balance you would do well to purchase such solutions ready-mixed, or have them mixed by a pharmacist.

6. The Structure of Plants

Introduction to the Hand Microtome Technique

You have probably observed, when you made the examinations of simple fresh preparations, that all organisms are composed of cells. In one-celled organisms, like the paramecium and the amoeba, the entire body consists of only one cell. In some algae you saw similarly-formed cells united in a group. Now you will find that higher plants are made up of many-cell groups—tissues—which have varied tasks to perform in a body and for this reason are of very different constructions. First, you should examine comparatively simple plants, such as the moulds.

MOULDS

You find moulds on rye bread, lemons, oranges and preserved fruits in profusion. They are simple plants without chlorophyll. You can get these moulds by a simple culture in a so-called damp chamber.

Make the chamber in this way. In a flat plate filled with water place a saucer with the bottom up. On this place a piece of moistened rye bread. Then put a jam jar or a glass bell over the saucer so that its edge rests in the water.

After only two to three days a mould will form on the bread. This will possibly be a whitish mould whose threads grow in a thick tangle. Transfer a little of this tangle with a lancet needle to a water drop on a slide and examine with medium magnification. From the white threads there grow stems with little heads (sporangia). This is a common bread mould (*Mucor mucedo*). The little heads disintegrate very easily and set free brown spores.

To make a permanent preparation transfer a small bundle of the mould threads directly to chrome acetic fluid (see fixing of diatoms), and fix it in this for 15 to 30 minutes. (Shake frequently.) Then wash in tap water with at least three changes (15 minutes); finally transfer to distilled water. For dyeing, use hemalum or Heidenhain's iron hematoxylin. Hemalum dyes rapidly, but the slower iron hematoxylin makes a more attractive preparation.

Dyeing with hemalum: Transfer the fixed tangle of mould from distilled water to Mayer's acid hemalum for about 10 minutes. Then wash it for at least 10 minutes in several changes of tap water. The mould threads which were first tinted reddish now turn a deep blue. Dehydration follows. You can put the moulds directly in methyl glycol (at least 10 minutes, longer does no harm), then in terpineol and xylol (5 minutes each) and finally mount in resin. If methyl glycol is not available, you have to use the alcohol stages: 35, 50, 70, 80, 90 and 95% alcohol, 5 minutes in each, terpineol and xylol for 5 minutes, then mount in resin.

In dyeing with Heidenhain's iron hematoxylin, you need a 2.5% iron alum solution and a hematoxylin solution. From light violet, fresh iron alum crystals, prepare a 10% stock solution in distilled water. To use, dilute a part of the stock solution so that a 2.5% solution results. To prepare the hematoxylin solution, dissolve 0.5 gm. hematoxylin in 10 cc. of 95% alcohol (not industrial spirit) and dilute it with 90 cc. of distilled water. Pour this hematoxylin solution into a small-necked bottle and cover the opening with a paper cap. Only after about four weeks is the hematoxylin mature and ready for use.

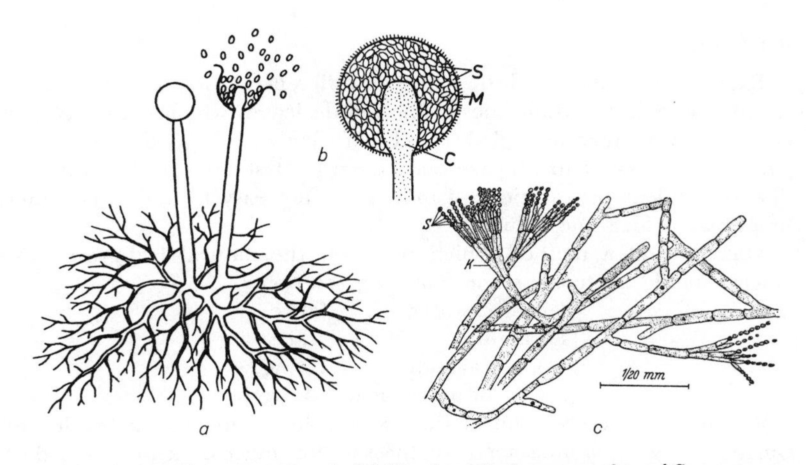

Fig. 70. (a) Bread fungus (*mucor*). **(b) Single-celled sporangium (S — spores, M — membrane, C — columella.) (c) Penicillium.**

The mature solution should be tightly sealed, and it keeps for a long time.

Put the fixed and washed mould threads for at least 2—better 12—hours in 2.5% iron alum solution; in this they are "tanned". Then thoroughly rinse them in distilled water and transfer to hematoxylin solution, which you have diluted before using in equal parts of distilled water. Let the moulds remain for several hours (to two days) in hematoxylin, which colors them deep black. The overdyed threads are rinsed in distilled water and "differentiated" in the 2.5% iron alum solution already used for tanning. As soon as the objects are in the iron alum, black clouds of dye arise: the hematoxylin is drawn out again. You must then interrupt this process at the proper moment, namely, when the individual structures stand out clearly. As you have not had much practice, check the progressive bleaching processes at short intervals: rinse the objects with water (which interrupts the differentiation) and examine microscopically. If the dye is still too strong, put them back into the iron alum solution again. When the proper dyeing has been attained, wash in frequent changes of tap water to eliminate all traces of iron alum. Finally dehydrate in methyl glycol or by alcohol stages, then terpineol and xylol and afterwards mount in resin. Good preparations show sharp contrasts in blue-black.

Heidenhain's iron hematoxylin is one of the most valuable dyeing methods in microscopy. Because of the differentiation possible with this method you can show the finest structural details of animal or plant cells with otherwise unattainable sharpness and clarity.

MOSS LEAVES

Examine in water the little leaves of different mosses, which are to be found everywhere. Many mosses have little leaves which consist of only one cell layer. Such a single-layered leaf (*Mnium* is a good one) is what you will next want to prepare permanently. It should of course be said that permanent preparations of mosses are not easy to make. You must be prepared for some failures.

Make use of a method which preserves the natural leaf-green. The objects should be fixed in the following solution:

Distilled water	90 cc.
Formalin	8 cc.
Copper acetate	10 mg.
Lactic or acetic acid	5 drops

You must use quite a lot of this fixing solution for only a few leaves. During fixing, the container with the solution must be kept in the dark in a cabinet or a paper box. After about 6 hours the fixing process is

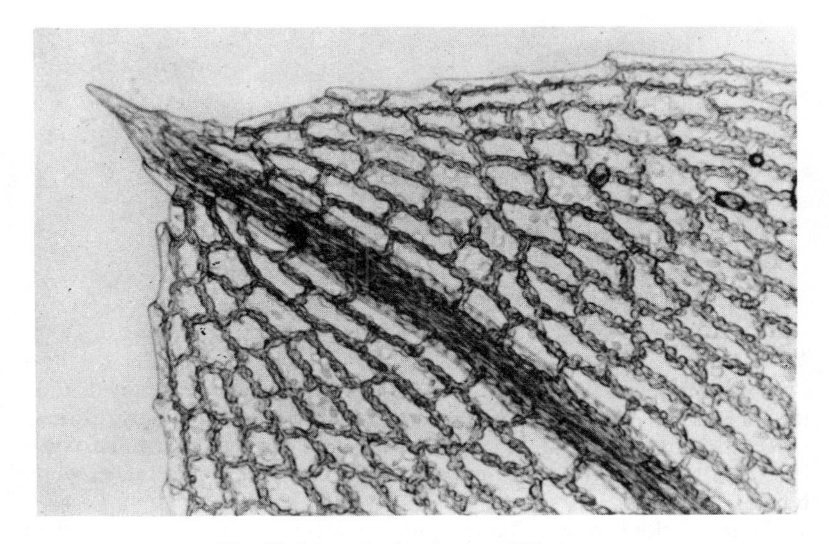

Fig. 71. Moss leaf enlarged 115 times.

ended. The moss leaves are then washed well in many changes of distilled water and mounted in glycerin jelly.

MALE FERN

Place a small piece of a frond with clusters of spores (sori) into a drop of water under the microscope and study the formation of the spore clusters. The piece of frond can be made into a permanent preparation as for moss leaves.

Fig. 73. Pinnulate leaf of a fern with sori.

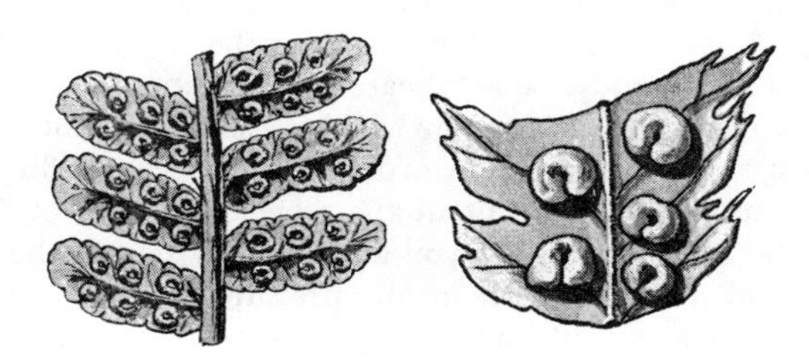

Fig. 72. Balsam bottle with cap and glass rod.

SECTIONING WITH A RAZOR

Many specimens, when gathered, are too thick to be examined under the microscope. With an ordinary microscope you can only examine objects which are so thin that light can pass through them. To try to look at a rose stem without any further preparation, for instance, would be quite senseless.

To study thick objects under the microscope you must make them thinner by some satisfactory method. The most useful procedure is the preparation of the thinnest possible cuts with a razor. The best razor to use is one which is especially manufactured for the purpose; it is flat on the underside and concavely ground on the upper side. Barbers' razors which are ground hollow on both sides are not as good. At a pinch, an unused heavy razor-blade placed in an available holder will serve instead of a cut-throat razor.

In cutting, lay both arms comfortably on the table. Hold the object in the first three fingers of the left hand and the razor in the right hand near the hinge. When cutting, let the back of the razor slide against the first and second joints of the left index finger, which gives a nice, easy guide for cutting. Always move the razor by pulling, never by pressing. The cut must be a "shear," not a "chop".

First cut off a somewhat thick slice from the object in order to prepare a smooth surface. In order to get a section for observation, you cannot simply begin to cut at the edge of the object. The severed slice would be much too thick; you do better by setting the razor on the cut surface and pulling it without any pressure over the cut surface. In this way the slice tapers off and is very thin in at least one place.

The biggest mistake you can make when cutting is to press the razor through the object instead of drawing it through. If possible, the razor should be drawn its entire length through the object; only in this way will you get thin, even sections. In cuts made under pressure the slices

Fig. 74. Start the knife at the right corner of the object to be sectioned and draw it forward slowly and without pressure along the entire length of the blade until the section can be lifted from the blade.

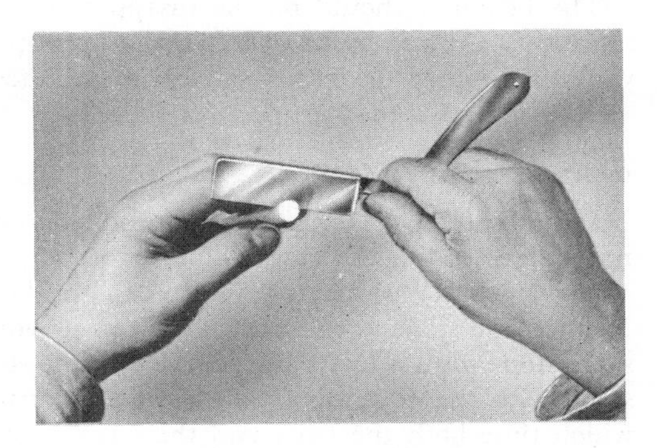

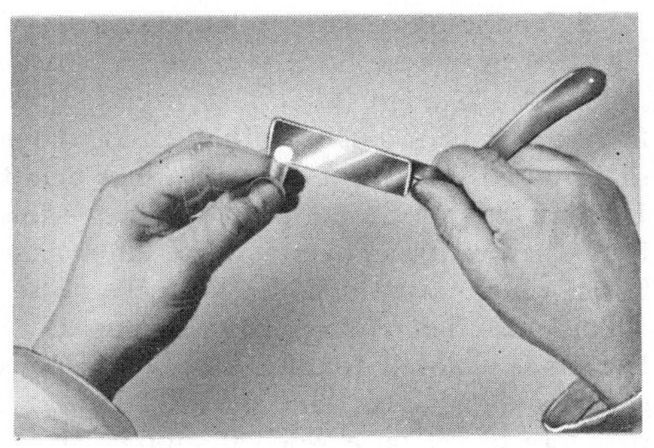

will not only be too thick but all the finer structures will be crushed and torn.

Another thing which must be emphasized is that a cut should be made toward, not away from, the cutter.

To hold the object, use elderberry pith. Place only flat objects between the halves of the split pith. For stems, twigs, etc., hollow the pith so that the object will fit in snugly when both halves are put together. Only in this way is it possible to make really thin sections and to make any sections at all of objects with large hollow spaces (grass stems) and loose pith (reeds). It is impossible to slice these freehand.[11] If you want to make sections which are also thin enough for finer examinations, you must usually forget about cutting across the entire surface of the object. Tiny, wisp-thin fragments often show more detail than cuts across the entire surface. Only for a gross survey is a thicker cut through the entire object useful; for finer examinations use thinner partial cuts.

The beginner should not be disappointed if the first sections are not successful. Cutting by hand requires a lot of practice. After several weeks you will become dexterous enough to produce thin sections.

CARE OF THE RAZOR

It is absolutely essential for the production of fine sections to learn the necessary steps in the care of the razor.

Using a blue "water stone", you can grind out little nicks very quickly. Moisten the stone with water and dress it with a broken-off piece of the same stone until a fine paste remains on the stone. Place the razor on the stone with the edge turned forward and turn it in little circles, during which time both the back and the edge must rest on the surface of the stone. You must be careful that both sides of the edge are evenly ground. The razor must be turned on its back, never on its edge. From time to time test the edge (after first wiping it off); it should "stick" to the fingertips when touched, that is, it should penetrate the outer skin slightly and should cut through a hair held free without pushing it aside. If this is the case, then draw the razor—this time with the back edge in front—several times over a strop. After grinding on the stone, the edge has a roughness which can scarcely be seen with the naked eye, but which must be removed with the strop. The ordinary strop used by barbers is not well adapted for microscopic work. Your razor should not get a rounded facet, but should retain its flat facet. The strop should therefore be set on a wooden support; it is better to use a square strop which has a stone (usually useless) on one side and leather on the other

[11] Elderberry pith is gathered in the autumn from dead shoots. They must not have become spoiled. If elderberry pith is not available, you can substitute a piece of cork. Cut the pieces of cork from a good, soft bottle cork having as few holes as possible in it. The small corks that pharmacists use are well suited for the purpose. The coarse corks from wine bottles, etc., are not usable. For cutting harder objects cork is greatly preferred to elderberry pith.

three. The sharpest leather is the red-colored one, which you use first, then the black and lastly the natural or white leather. Occasionally, you should rub the strops with butter or olive oil to remove the old coating of strop dressing. Then apply new strop dressing, which is best applied to the leather with your fingers or a ball held in your hand.

Fig. 75. Four-sided strop or hone.

Stropping is done by pulling the razor diagonally forward with the back in front, so that with each stroke the entire surface is drawn over it. At the end of the strop turn the razor on its back and draw it back in the opposite manner. Stropping too long on the black and on the unfinished side should be avoided. Otherwise the edge will become too "flat" and not "bite" properly when cutting (Fig. 76).

It is very important that the razor is stropped not only before using, but above all after using. The edge shows teeth like a saw, visible under the microscope. When you dry the razor, you can never take care of all the tiny drops of water which cling between the teeth. These little drops of water form rust, which shows up the next time the edge is used. How-

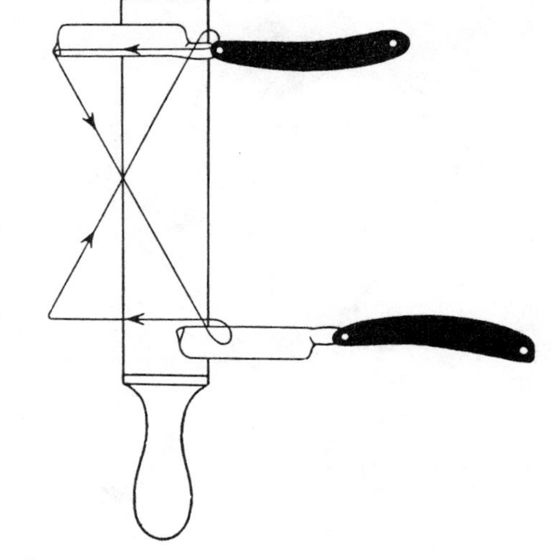

Fig. 76. Honing the razor on the square strop.

Fig. 77. Edge of razor, greatly magnified; (above) after honing on the stone; (below) same edge after stropping.

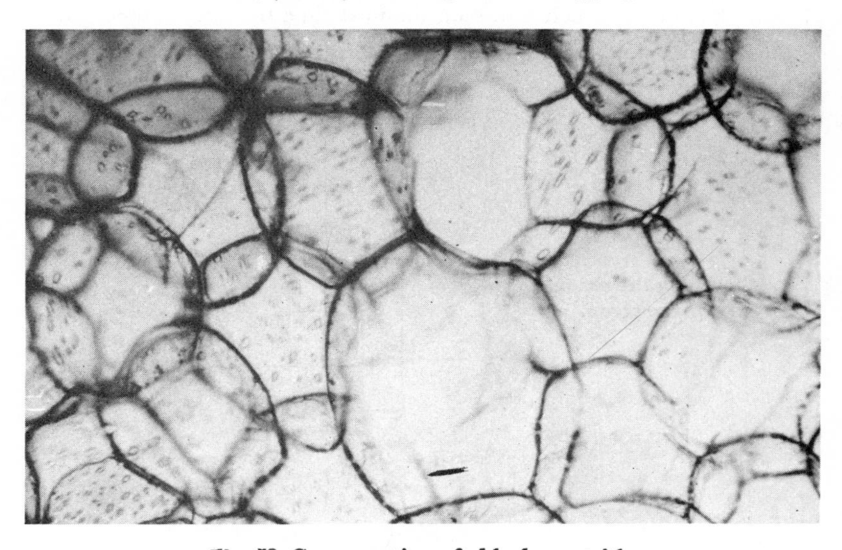

Fig. 78. Cross-section of elderberry pith.

ever, if you draw the razor after use again over the red side of the strop, the strop paste absorbs the moisture and covers the edge with a protective oily film.

Never allow the razor to remain open on the work table. Not only do you place yourself in danger of injury, but the very sensitive edge suffers also.

CORNSTALKS (*Zea mays*, Maize)

Cut pieces 2 to 3 cm. long from a cornstalk which has been split, and place them for fixing and hardening in spirit. (Spirit itself is a very poor fixing medium; since the cell contents will later be removed, you can make an exception here and use it.) The longer the material lies in the alcohol, the easier it will be to cut. You can try cutting it after a few days, but even weeks in the high-percentage alcohol does no harm.

The pieces of stalk which have been hardened in the spirit are clamped between pieces of elderberry pith and you then attempt to make very thin slices according to the instructions given above. Sometimes it is better to moisten the razor with a drop of alcohol. You must be careful however that the pith always remains dry; damp pith is much too soft for cutting.

No particular measure can be given for the thickness of hand cuts. You can make this clear to yourself with an example of how a perfect cutting should look. For this use the elderberry pith cuttings made in practice. They are usable if they show only one layer of tissue in the cross-section of the pulp. If the lower cell layers still glisten through, the cutting is too thick. The best hand cuttings are approximately 30 microns (one micron equals 1/1000 mm.); in the most successful instance 20 microns may be achieved when a cut "tapers" out. In many cases, cuts which are 50 microns in thickness are still usable.[12]

First make an overall cut, going through the entire organism from which you will take sections. This slice can of course be thicker, as it merely serves as a rough orientation for locating the tissues (see Fig. 82). The finer structures are disclosed by more delicate cuts. If a stem is too thick,

[12] If you place particular value on cuttings of accurately determined thickness you must avail yourself of one of the common hand microtomes. These simple, not too expensive, instruments consist of a cylindrical tube which has a ring-shaped insert on its upper end in which is set a glass plate of 70 mm. diameter. A second cylinder slides in the first; it can be moved up and down by means of the micrometer screw seen below. The thread of the micrometer screw is on the outside edge. Every mark shows a lifting of the inner cylinder by 1/100 mm. The object to be sectioned is fastened in the inner cylinder by a set-screw on the side. When cutting, the microtome is held with the left hand, while the right draws the razor through the object, guided by the glass plate; the razor must for this reason lie firmly on the glass plate through the entire cutting operation. This simple apparatus is naturally unusable for the finest sections; for this purpose a larger microtome can be used. (See Fig. 79.)

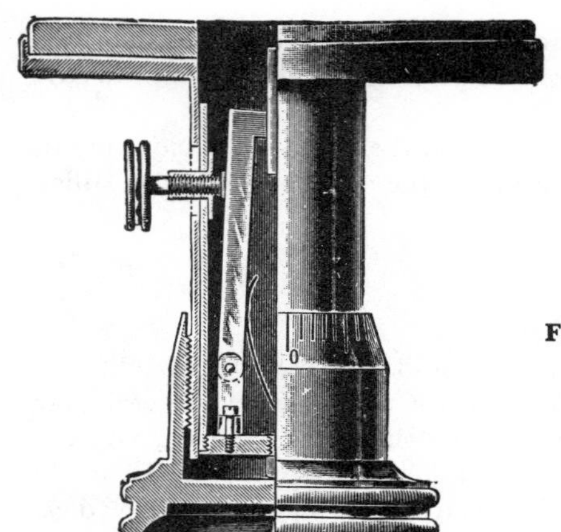

Fig. 79. A simple hand microtome.

it is halved, quartered, etc. It is always advisable to mount it in elderberry pith or cork.

To maintain a constant cutting direction is absolutely essential. Only a few cells and structures show the same microscopic view in sections made in different directions. Fig. 80 shows the different cutting directions in a stem. In cylindrical organs (as, for instance, in the cornstalk), the transverse section runs across, and the longitudinal section runs parallel with the axis of the organ. While the cross-section always shows similar structures, a longitudinal section can show varying features. For this reason you must differentiate between the radial longitudinal section, which meets the radius of the cross-section, and the tangential longitudinal section in which the surface of the section is vertical to the old radius. In order to be able to demonstrate the structure of a plant stem, you must examine sections in all three directions.

Always make a great number of sections (perhaps 15 or 20) which you can sort out successively in water. Those which prove to be too thick can be discarded.

The finished sections are removed from the razor with a soft brush and are transferred to a watch-glass filled with water. For a preliminary examination, lay a number of sections on a slide in a drop of water, cover them with the coverslip and sort them with weak magnification. Remember the positions of the sections which are judged to be too thick, and discard them immediately. In order to remove the coverslip without damaging the delicate sections, put enough water at the edge so that the coverslip begins to float. With the help of pointed tweezers and a preparation needle it can then be easily removed. The successful sections are transferred to another watch-glass with water, and the bad ones discarded.

For further preparation the sections must now be placed in several solutions; you must dye them, dehydrate them and finally mount them in resin. Transferring the sections from one fluid into another is not exactly simple. You can make use of several methods:

1. The solutions are poured into different watch-glasses and the sections are transferred from one to the other with the lancet needle or better yet with a fine brush. The method has the disadvantage that the sections are easily damaged and often get lost—especially in opaque dye solutions.

2. The sections remain in the same watch-glasses throughout all stages of preparation. The fluid is drawn out of the glass with a finely-drawn-out pipette and is then replaced with the next fluid. In this procedure the danger exists that the sections will dry out or that so much fluid is left behind in the watch-glass it makes the next impure. In addition you must—to avoid heavy impurities—work with a number of pipettes. You need at least one pipette for alkalis and acids, one for dye solutions, one for the dilutions of the alcohol stages and one for the non-aqueous fluids such as alcohol, terpineol or xylol. The pipettes must never be mixed up under any circumstances.

3. The most elegant method is to handle the sections with a dyeing sieve. You can easily make such a sieve. Take an ordinary test tube and file off the top 2 to 3 cm. with a glass file or an ampule file (your pharmacist or doctor might make you a present of an old ampule file). You

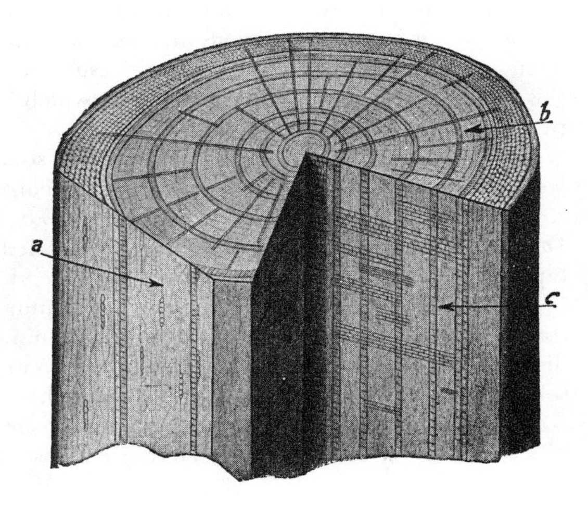

Fig. 80. The different cutting directions with cylindrical objects. (a) Tangential cut. (b) Cross cut. (c) Longitudinal cut through a wood stem.

now have a tube which is open at both ends and one end (the upper) flared a bit. Now cover the upper end with a piece of old nylon stocking using thin copper or brass wire tightly wrapped around to hold it stretched in place. The surplus ends of the material are trimmed off. When using such a dyeing sieve place the individual fluids in small, cork-sealed preparation jars. Lay the sections in the sieve and place the sieve in the preparation jar which has the required fluid. If the fluids must be changed, the entire sieve is removed, the solution is allowed to drip off and then it is placed in the next fluid. Almost all the disadvantages of the previous two methods are avoided. This sieve should not be placed in strong acids or alkalis (also not in Javelle water). If it is necessary to place the sections in Javelle water you must take them out of the sieve, place them in a watch-glass with the Javelle water and after washing them off, place them again in the dyeing sieve.

Your first task after cutting was to sort out the good sections under the microscope, selecting those which seemed suited for permanent preparation. Now place the sections in Javelle water (bleach), in which the cytoplasmic cell content is destroyed, leaving only the cell walls. The result is an uncommonly clear, easily viewed image in the microscope. The sections remain in the bleach until they are all white, 10 to 30 minutes depending on their solidity. They are then lifted out with the lancet needle and placed in a watch-glass with distilled water, which you have previously weakly acidified with a trace of acetic acid (one drop of acetic acid to about 10 cc. of distilled water). After several minutes, transfer the sections in a similar manner (or with the dyeing sieve) to pure distilled water and now begin to dye.

For dyeing botanical sections the use of hematoxylin-safranin dye is often recommended. It produces nice results, but takes quite long. For this reason a short-cut is described below, which lets you get a permanent preparation in the shortest time.

The hematoxylin-safranin dye is a double stain, which dyes the non-woody cell walls blue and the woody ones (in corn, especially the vascular structure) a bright red. The sections are dyed for 5 to 10 minutes in Delafield's hematoxylin, briefly rinsed in distilled water and "blued" in tap water. If the sections which at first were reddish have turned a pretty blue, put them for contrast dyeing in a safranin solution in which they stay for 2 to 8 hours. (Composition of the safranin solution: 2 gm. safranin dissolved in 100 cc. of 50% alcohol. In dyeing, 20 cc. of this stock solution is diluted with 80 cc. of 50% alcohol.)

The bright red dyed sections are quickly washed in distilled water and differentiated in fuel alcohol. Here they lose their red color again.

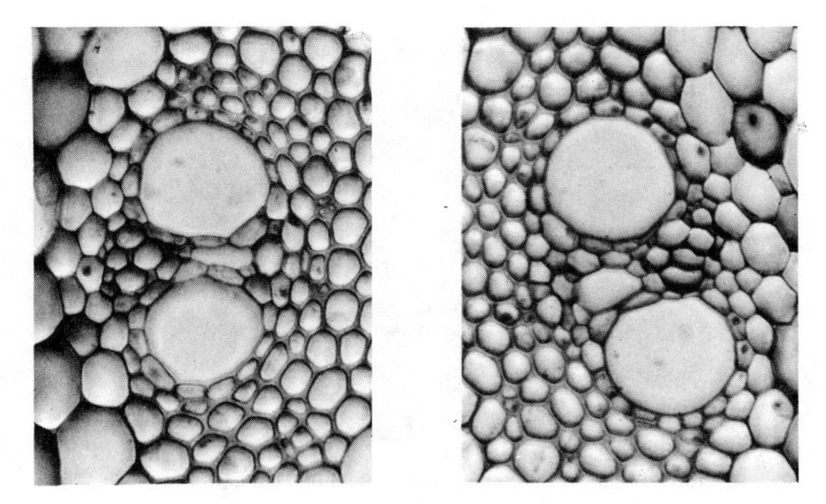

Fig. 81. Vascular bundles of corn, from a cross-section of the stalk (relatively thick hand cut) magnified about 100 times.

Differentiate until only the woody cell walls show a red color; the non-woody cell walls again show the blue tone of the hematoxylin dye. To interrupt the differentiation, transfer the sections for several minutes to absolute isopropyl alcohol, which at the same time removes all traces of water, and then transfer them to terpineol. The sections may remain a long time in the terpineol, if desired. Leave them in for at least 5 minutes, and then as always mount in resin.

The contrast dyeing with safranin can be eliminated. Good results also can be attained with the use of hematoxylin dyeing alone; hematoxylin is above all a universal dye: in all cases where in doubt about a proper dyeing method, try a hematoxylin solution. There is hardly a single object—whether it be of plant or animal origin—which cannot be clearly and cleanly dyed with hematoxylin. You can often get particularly sharp dyeing by using Delafield's hematoxylin strongly diluted with distilled water, and leaving the objects in for a longer period of time. (The weaker the solution, the longer the period of dyeing by hours, even days.) Every beginner is advised to place his trust in the hematoxylin dyes and to use them extensively; they belong to the most reliable, trustworthy and time-tested dyeing methods.

You get very pretty tints in plant specimens by using direct deep black. Prepare a saturated solution in 70% alcohol. Before dyeing, the solution

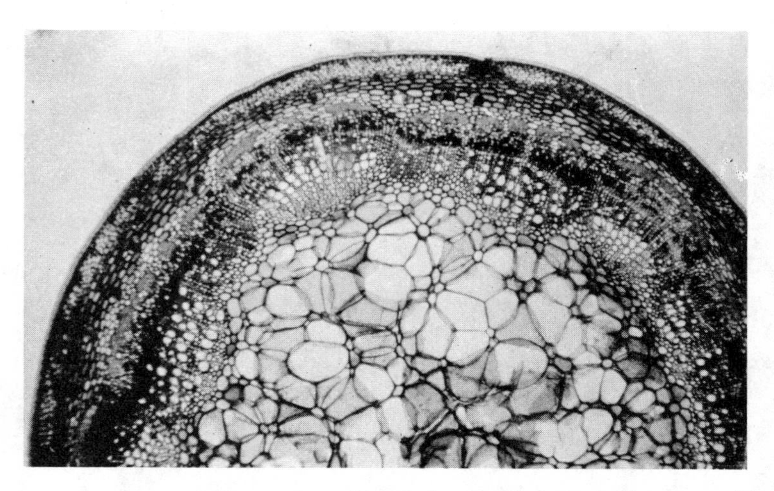

Fig. 82. Rose stalk, cross-sectioned for rough orientation, magnified about 40 times.

is filtered. The sections are placed in the dye solution for 5 to 30 minutes. (Length of dyeing depends on the thickness of the section; thicker sections take dye very rapidly, thinner ones take longer. Microtome sections must sometimes be left in the dye for hours.) After dyeing wash in 90% alcohol, until no clouds of dye are given off. If the sections are still too pale, put them back into the dye solution. Before re-dyeing remember that a differentiation is extremely difficult. After the 90% alcohol there follow —as described above—absolute isopropyl alcohol, terpineol and resin.

A short-cut which works very reliably, leaving the non-woody cells grey-black and the woody ones yellow, is dyeing with picro-nigrosin. Combine 10 cc. of 1% aqueous nigrosin solution with 90 cc. of saturated aqueous picric acid solution. (Dry picric acid is explosive; for this reason use it in solution.) The dye mixture can be stored. Take the sections which have been bleached with eau de Javelle, wash again and place them for about 15 minutes in picro-nigrosin. (Length of dyeing period depends on the cut and the nature of the object; under certain circumstances it must be changed somewhat.) From the dye solution, transfer the sections directly to 90% fuel alcohol; then after a few seconds to a minute at the most in 100% isopropyl alcohol which you allow to work in for 5 minutes. During the latter period change at least once. From the 100% isopropyl alcohol the sections are placed in terpineol (5 minutes), then mounted in resin. The grey-black tones often are somewhat pale with this procedure, but at the same time the finer structures are not "smeared", as with some more colorful dyes.

In this dyeing—just as with safranin dyeing—you cannot use the alcohol stages—because alcohol which contains water (here, in the picric acid) would again remove the dyes. The long step from water to high percentage alcohol is not undergone by the delicate sections without considerable structural change and shrinkage. With the help of a little dodge you can however largely prevent the shrinkage: place the dyed sections on a slide in a drop of dye solution. Absorb the dye solution with a bit of filter paper until the sections are dry and tightly stuck to the slide. In no case, however, may the sections be allowed to dry out! They should be just a little damp. With a dry pipette, drop 90% alcohol on the section. The section then floats in the drop of alcohol, is hardened at the same time by the alcohol and may easily be transferred to a dish of 100% alcohol with a dry or alcohol-washed brush. The same method is also helpful when larger, thin slices stick together as they are transferred in alcohol.

If you want to make a permanent preparation quickly, put undyed sections directly from water into a mount of glycerin or glycerin jelly. Such a preparation cannot be compared to preparations which have been made as described above.

CLEANING PREPARATIONS

If you do a clean job you should have no trouble keeping your preparations clean. The resin layer should only be heavy enough to just fill out the space underneath the cover-glass. If, however, some resin oozes from out under the cover-glass, you must never remove the surplus with xylol, because xylol will dissolve the resin, get under the cover-glass and smear the whole preparation. Wait patiently until the resin is hardened somewhat and then remove the surplus mechanically with a knife. The last traces are removed with a benzine-moistened cloth.

IRIS LEAF

The iris is easy to procure. If you cannot get an iris leaf, then you might examine a tulip leaf, leek leaf (Fig. 83) or lily-of-the-valley leaf. For a preparation of the cross-section, a gladiolus leaf is also very good (Fig. 84).

First complete a plain preparation of the upper epidermis of the leaf. To do this, stretch the leaf with the thumb and middle finger over the index finger of your left hand and make a very light cut with the razor.

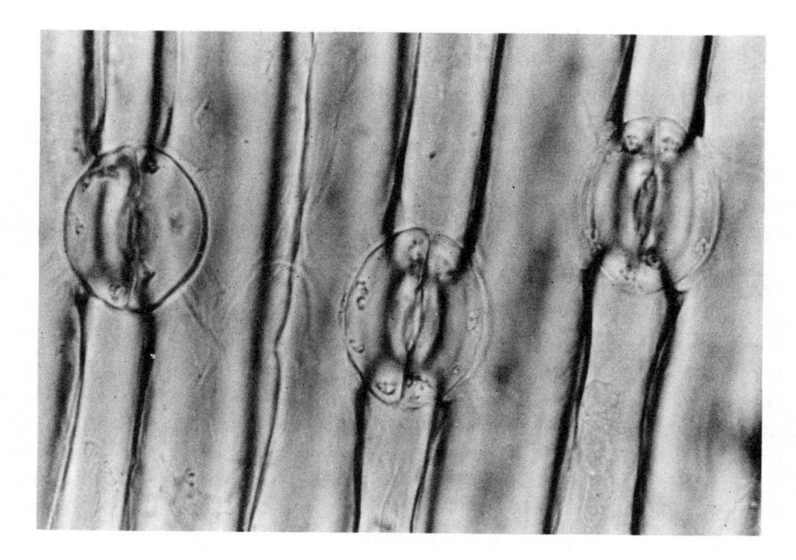

Fig. 83. Outer skin of a leek.

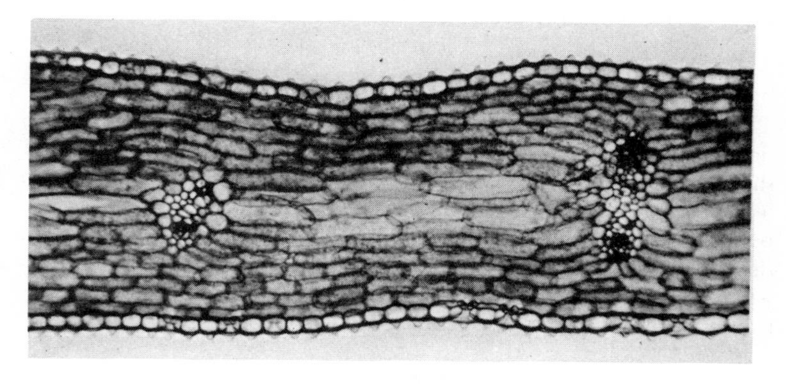

Fig. 84. Cross-section through the leaf of a gladiolus (relatively thick hand cut, magnified about 450 times).

The edge of the cut is grasped with the tweezers and a bit of the upper skin is pulled off. The skin which is removed must be colorless; if it appears green, it is too thick for microscopic examination. Put a small bit of the skin in a drop of water under the microscope. Note the colorless,

Fig. 85. Clamping a bit of leaf in elderberry pith.

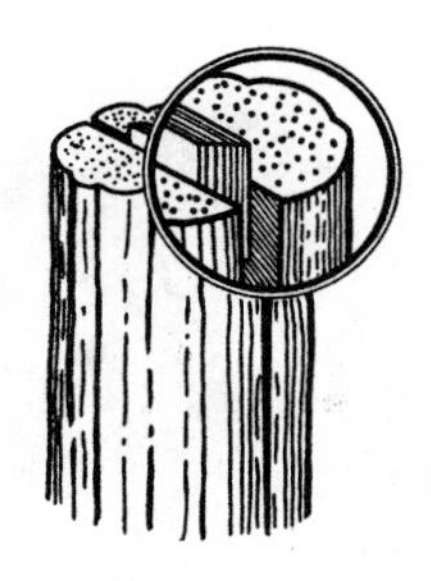

elongated cells and stomata of the upper epidermis. The stomata facilitate the gaseous exchanges of the plant. For completion of a permanent preparation, as for corn stalks, a simple hematoxylin or direct deep-black dyeing is satisfactory here. Before bleaching in Javelle water a treatment with alcohol for several hours is advisable because only then is the cell content completely removed.

A cross-section shows not only the finer structure of the stomatal mechanism, but the entire inner structure of the leaf. Along the midrib, cut out a narrow strip, at most $\frac{1}{2}$ cm. wide. Fold this strip over and over, then clamp it in elderberry pith and try to get some very thin sections. Because the leaf has been folded, you get a greater cutting surface and in addition get a greater number of sections with each cut. Subsequent stages are as for cornstalks.

ORANGE AND LEMON PEEL

Handle orange and lemon peel in the same manner as for cornstalks, hardening in alcohol, prepare thin sections and dye with hematoxylin or direct deep black. Before dyeing, examine the sections in water and glycerin.

In the outer part of the peel you can see large and small holes, which were filled with volatile oil. These holes contain the oil glands. You know that an orange peel gives off a sweet-smelling fluid if it is squeezed. The pressure bursts these oil glands, the form and structure of which you now can study exactly. Of course, you can no longer see the oil in your preparations; it was dissolved by the alcohol.

NUCLEUS DIVISIONS AND GIANT CHROMOSOMES

Place an onion in a glass filled with water, so that the end of the onion just touches the surface of the water. After a few days the onion will have put out quite a number of roots. Cut off such a root (this is best

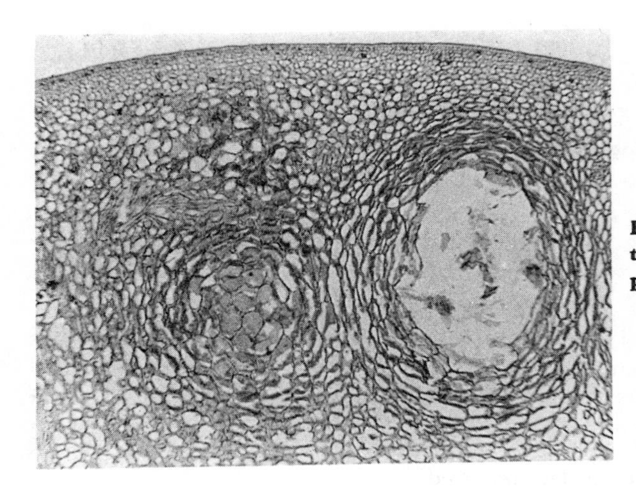

Fig. 86. Cross-section of the outer part of a lemon peel.

done in the early morning because then the greatest number of divisions are to be found). Cut the fresh root in half with a razor. The first millimeters of the root tip are the most important; further back, in the older root section scarcely any more nuclear divisions occur. The halves of the root tips are fixed in a mixture of 3 parts of absolute alcohol (here you can also use isopropyl alcohol) and one part of acetic acid for 15 minutes.

The root tips are then transferred directly to a small test tube of acetocarmine[13] and boiled in it for 2 minutes. Transfer the root tips from the hot dye solution to a slide in a drop of fresh acetocarmine and place a coverslip on it. By a downward pressure on the coverslip with a needle the root tips are squashed so flat that only one cell layer is to be seen under the microscope. Microscopic examination reveals the chromosomes of the dividing nuclei as a bright red on a light background. If the color is not bright enough, you can boil the dye solution again and again under the coverslip (by holding the slide over a flame). Sometimes it is necessary to add more dye solution at the edge of the coverslip.

At first, the tint will always be more intense, but it will eventually lose its sharpness. Good permanent preparations cannot be produced with this dye. If you do not want to draw the division stages which are very easily detected, you may—if you have the equipment—take micro-

[13] To make the acetocarmine solution: 45 cc. of acetic acid and 55 cc. of distilled water are mixed in a boiling flask or an Erlenmeyer flask and 4 to 5 gm. of pulverized carmine are dissolved in it. After setting up a reflux condenser or a long distillation tube, allow it to boil over a small flame for half to one hour. The fully cooled solution is filtered and if well sealed can be kept indefinitely.

photographs (see page 140). The root tips of sprouting beans are also well suited for the examination of nuclear division.

With some luck you can, with the help of the acetocarmine method, gather all division stages of the cell nuclei. For a permanent preparation hematoxylin is better, particularly Heidenhain's hematoxylin. Microtome sections, which should not be thicker than 5 microns, are a necessity.

The giant chromosomes in the salivary glands of fly and gnat larvae are famous because of their great importance in genetic studies. You can easily examine them with the acetocarmine method.

In wintertime you can buy gnat larvae very cheaply in any aquarium supply store. Remove the head of one of these larvae with a razor blade and squeeze it a little with a lancet needle. From the surface of the cut will appear two watery bubbles, the salivary glands, which you cover with a cover-glass and examine. Even under a weak magnification you can see the large cell nuclei of the salivary glands, in which the giant chromosomes lie. These are twisted, worm-shaped structures, which show a characteristic banding. For these giant chromosomes, dyeing with acetocarmine is also very useful. The bands, in which are found the hereditary substances, the genes (genetic factors), take on an intensely red color. It is advisable to make examinations of acetocarmine dyed preparations with a green light filter; the chromosomes then appear almost black and as sharp as if they had been dyed with iron hematoxylin.

STORAGE OF FIXED PLANT MATERIALS

There are many interesting objects that you may find on occasion, with which, because of the lack of time, you cannot immediately work. You must therefore store them in an appropriate fluid.

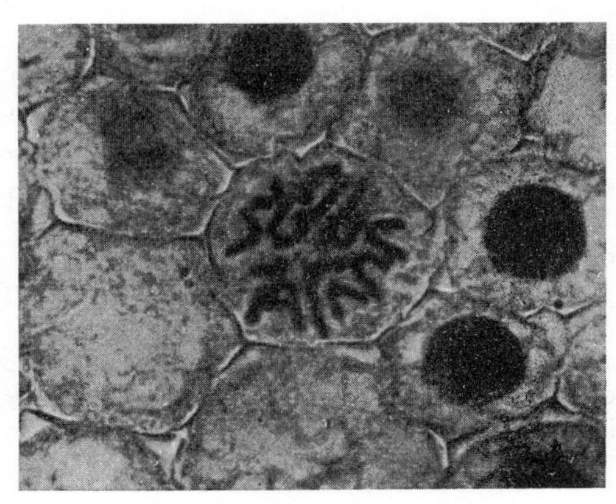

Fig. 87. Nuclear division in the root tip of soya bean. Medium magnification.

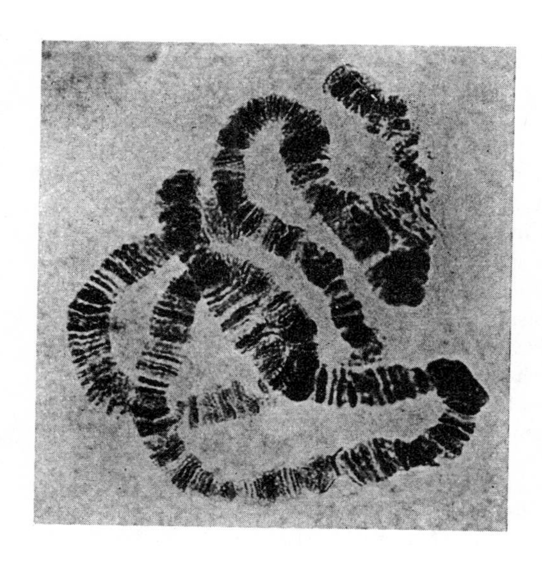

Fig. 88. Giant chromosomes from the salivary gland of a fly.

Every living object must first be fixed. There are a great number of different fixing methods, many of which are adapted for very specific purpose. Industrial spirit or absolute alcohol, with which you fix cornstalks, are very poor fixing substances, which you can use only for firm objects in a test preparation. On the other hand there is formalin-acetic-alcohol, a fixing solution where fuel alcohol may be used without damage, and which gives good results with most plant objects.

Formalin-acetic-alcohol is made up in the following manner:

Fuel alcohol	100 cc.
Formalin	10 cc.
Acetic acid	3 cc.

It is advisable to mix the individual ingredients of this fixing solution shortly before use. The diameter of objects to be fixed should not exceed 1 to 2 cm. Always take at least 50 times the mass of fixing fluid to the object to be fixed.

Fixing time is 24 hours. Then wash in 60% alcohol, which is changed 2 to 3 times in a 24-hour period.

To store fixed material use, in the case of plant tissue, a storage solution consisting of equal parts of alcohol, glycerin and water. You can use isopropyl alcohol but never industrial spirit.

The objects are transferred from the 60 to 70% alcohol to the storage solution in which they may be preserved for decades. An important thing, of course, is that the storage jars be well sealed. For storage containers, large preparation jars are well suited. Because alcohol evaporates even with perfectly fitted corks, it is better when storing for a longer period to use bottles with fitted glass stoppers. Use a large glass-stoppered jar with the biggest mouth available. This is filled with the storage solution. In addition, buy a number of small test tubes. Place the fixed objects in one of the little test tubes, fill with the storage solution, plug with a little alcohol-soaked wad of cotton and then stand them in the large storage jar. If you have chosen a large enough jar, you can get a great many of these test tubes in it and in the course of years get together quite a collection of materials. Naturally every individual bit of material must be identified: insert a little note in the test tube—before the cotton plug is inserted—on which you write in pencil or ink the nature of the object, where found, and the date and method of fixing.

If in the course of time the storage solution should begin to evaporate, an equal amount of 70% alcohol is added. After a storage of three years the solution in the storage jar should be renewed. Before working on stored materials, wash them off in 70% alcohol and if so instructed harden in 95% alcohol.

HANDLING FRESH SECTIONS

It was noted for cornstalks that usually any plant material is easier to cut after alcohol hardening than in a fresh condition. But the alcohol hardening takes much time; also in many cases it is necessary to examine the living material. If you have got experience in cutting alcohol-hardened materials, you should also practice with living plant parts.

Sections of living plant material are immediately examined in tap water. If you like a section particularly well, or have found an interesting subject on a section, it is still possible to make a permanent preparation of it.

For fixing sections the chrome acetic fluid is particularly well adapted (page 70). Fixing time is 10 to 20 minutes. After washing the fixed materials, dye them as previously instructed. Particularly fine stains can also be got in higher plants after fixing with chrome acetic fluid by the use of alizarin-viridin chrome alum (page 72). Very good results can be got with alizarin-cyanin RR: in 100 cc. of distilled water boil 0.25 gm. of alizarin-cyanin RR and 5 gm. pure aluminium chloride. Filter after cooling; after 8 days repeat the filtration. The sections remain in alizarin-cyanin for about a day; then they are washed thoroughly in distilled water and dehydrated in the usual manner. Mount in resin.

DETECTION OF FAT

For fat detection you need an alcoholic Sudan red solution which is prepared in the following manner:

Heat 50 cc. of 50% isopropyl alcohol (equal parts of distilled water and absolute isopropyl alcohol) on a water-bath (do not use an open flame). In the hot alcohol dissolve 0.1 to 0.2 gm. Sudan III. The cooled solution must be well sealed.

Make thin cross-sections of the kernel of a walnut or hazelnut. You must forget about previous hardening and fixing; the cell structure is easily cut without hardening. The cross-sections are placed without further handling in the Sudan red solution and left there for 15 to 30 minutes. They are then washed in distilled water and examined in water. For comparison examine other undyed sections in the water.

In the sections which have been dyed with Sudan III note orange-colored smaller and larger droplets in the cells. These are droplets of oil. Sudan III dyes all sorts of fatty substances. Technically, this is not a stain in the usual sense, but an absorption of the dye solution in the oil droplets: Sudan red travels from the poorer solvent substance—alcohol—to the better solvent—oil. Dyeing with Sudan red is a very sensitive microchemical reaction in fats and fatty substances. You can make permanent preparations of the Sudan red dyed sections. A mount in resin is out of the question, because the oil would dissolve in the resin. On the other hand a mount in glycerin jelly is possible.

DETECTION OF VITAMIN C

In 20 cc. of 1% acetic acid, dissolve 1 gm. of silver nitrate. This solution may be kept for a long time in a brown bottle. It is better to keep it in the dark.

Make thin sections across the green portions of a pepper or through parsley leaves or pine needles, and do not fix them. The sections are placed directly on a slide in a not too small drop of the silver nitrate solution; after a few minutes a coverslip is placed over it and the sections are examined in the silver nitrate solution.

After only a short time note the little grains of black in many of the cells—silver which has been deposited out of the solution. Wherever there is a grain of silver, there was formerly Vitamin C; because Vitamin C exerts such a high power of reduction that it can extract silver from a silver nitrate solution.

You find especially that the chlorophyll granules are strongly tinted black. You can now get a good idea through the distribution of the black grains of the distribution of Vitamin C in the living plant organism, and if you examine different portions of a plant in the same manner (leaves, stems, roots) then you can get important information about the distribution of Vitamin C throughout the entire plant. It is very intriguing to examine the Vitamin C content of various fruits and other food substances in this manner.

You can make a permanent preparation of these sections including the silver particles. For this purpose, take them from the silver nitrate solution and place them in distilled water, which must be changed repeatedly over a 10 minute period. From the distilled water they go for 10 minutes into a 5% sodium thiosulphate solution, and from this again for 10 minutes in frequently changed distilled water. Dyeing is accomplished in short order with eosin in 0.1% aqueous solution. The light pink sections are rinsed with distilled water. Then dehydrate either without stages in methyl glycol or in alcohol stages, finally in terpineol or xylol and then mount in resin.

7. The Microscopic Structure of Animals

The microscopic examination of cells, tissues and organs of which an animal is composed, is one of the most interesting of all studies. Of course, to make the finest examinations, microtome sections are necessary. An introduction to the use of the microtome will be given in the appendix. But even without a microtome the microscope hobbyist has such a broad scope of work available, that here only the most important lines of work can be indicated. Tissue structure and cell studies have already been shown in the chapter called "Exploring a Drop of Water" which dealt with the one-celled animals and plants. Now you will see that the cell is also the building-block of the higher Animal Kingdom as you have already found it to be in plants.

SNAIL TONGUES (*Radula*)

Kill a land snail quickly and painlessly by placing the animal on a flat surface, and severing its head with a quick stroke of the razor. The severed head is then boiled for 10 minutes in a 30% caustic potash solution, so that you can separate the soft parts and the radula. Boiling separates the soft parts and the fluid which was clear at first takes on a darker, brownish tint. After boiling, note the radula as a transparent bit of skin in the test tube. Pick up the radula and place it in a container of water, in which it remains for an hour with frequent changes. Then transfer to another container with a few drops of Grenacher's borax carmine. The time required for dyeing varies and tests must be made. For differentiation transfer the dyed radula into acetic alcohol solution, then to 90%, and finally 95% alcohol. To get out the last traces of water the radula is taken from the 95% alcohol and placed for 15 to 30 minutes in methyl benzoate. For mounting, first place a small drop of resin on the slide and lay the radula in it. When doing this, you must be careful that the many pointed chitin teeth which appear on the radula in orderly horizontal and vertical rows are on top. (Check with the microscope, using low magnification.) Sometimes it is necessary to spread out the curling edges of the radula with a preparation needle. Great care must be taken because the brittle skin tears easily when handled roughly. When everything is in order, place another drop of resin on top and cover with a coverslip (Fig. 89).

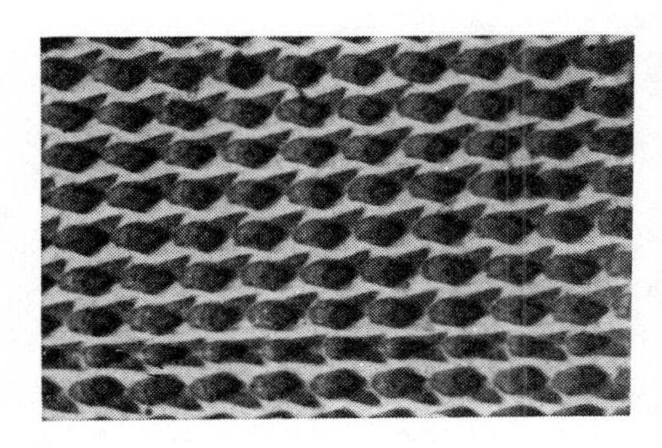

Fig. 89. A bit of rasping tongue (radula) of a land snail (Helix *pomatia*).

LIVER FLUKES (*Fasciola hepatica*—large liver fluke; *Dicrocelium lanceatum*—small liver fluke)

Obtain these parasites, which live in the bile ducts of sheep and cattle, from the slaughterhouse. The flukes are fixed by placing them in a formalin 1:4 solution (one part 40% formalin solution to 4 parts of tap water). Fixing time should be at least 12 hours; longer does no harm. Wash with 50% alcohol, which is changed 2 or 3 times in the course of a day. Before dyeing, transfer the liver flukes for a few more hours to 70% alcohol and then dye them in Grenacher's borax carmine for 1 to 3 days. For differentiation the objects are then placed for 2 days—better longer—in acetic alcohol (see page 102). After differentiation, washing takes place in 70% alcohol (the traces of acid must be got rid of, or your preparation will spoil in short order), and then dehydrated in 95% alcohol for 12 to 24 hours. From the 95% alcohol transfer the flukes to methyl benzoate: place the methyl benzoate in two small preparation jars. They remain in the first until they have sunk to the bottom; then they go into the second jar for 12 to 24 hours and are finally mounted in resin. Because these flukes are too large to be covered just with a coverslip, you must insert little pieces of broken slides.

On the leaf-shaped flat liver fluke you first see an oral sucker at the tip of the spindle-shaped anterior end of the body. Further back there is

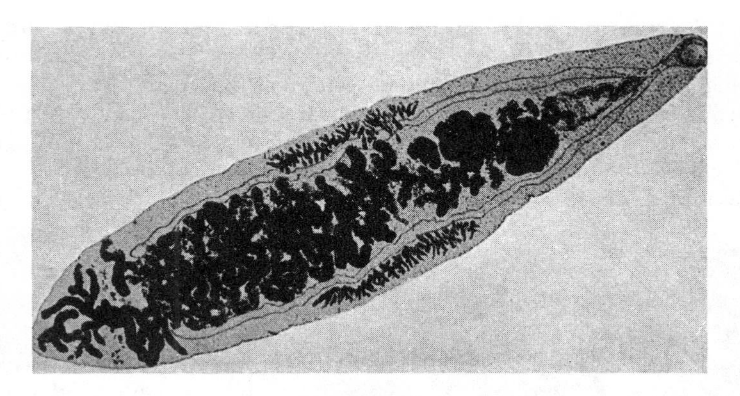

Fig. 90. The small liver fluke or leech (*Dicrocollium lanceatum*).

a ventral sucker. The endless forked intestine in the larger fluke has finely-branched blind pockets. Because the liver fluke is bisexual, it has testes, ovaries, yolks and a uterus.

You can prepare all animals of comparable size in the same manner as the liver fluke. Particularly suitable as an example of a vertebrate animal is a preparation of a little fish. Newly-born fry of the well-known aquarium fish, the guppy (*Lebistes*) are always available from your aquarist friends and aquarium supply dealers; they are easily prepared and make very interesting preparations.

FISH SCALES

Here differentiate between cycloid or round scales and ctenoid or toothed scales. Cycloid scales, as you will find on the herring or pike, are round discs; ctenoid scales (flounders) have a toothed edge on the free back edge.

The isolated scales can be placed in glycerin and there examined. To make a permanent preparation, boil the scales in caustic potash solution, wash thoroughly and mount in glycerin jelly or in resin. To mount in resin, it is of course necessary to dehydrate them: 70% alcohol half an hour, 95% alcohol 1 to 2 hours, methyl benzoate 2 to 12 hours, then resin. If the scales curl up during dehydration the slide and coverslip are clamped together with a clothes-pin and left that way until the resin has hardened.

A living fish—whether it be a little carp, minnow or stickleback, or maybe a goldfish (some aquarium supply stores sell off-color fish cheaply) —is easy to get as an examination specimen. Place the fish in a cloth and kill by hitting it on the head with a piece of wood. The tail region is carefully dried and the caudal base cut off with a pair of scissors. The blood which flows out is dropped on a slide which has been thoroughly cleaned with alcohol. Do a complete smear preparation (see page 49). It is important that you work rapidly because fish blood spoils very quickly.

Allow the smear to dry in the air. It cannot be treated further until after 2 hours, but in the meantime you can look at it undyed.

For dyeing make use of Wright's blood stain which is also useful for blood smears of other vertebrates including man. Buy Wright's blood stain ready-mixed. It can be stored for a long time, but must be very tightly sealed at all times.

Before dyeing, boil about 50 cc. of distilled water. It is best to use a pyrex glass container because a metal utensil could put impurities in the water. Allow the water to cool before using (see below).

The blood smear is placed in a flat glass dish (the cover of a Petri dish or a preserve jar). Drip on undiluted dye solution from a clean pipette. Allow 4 or 5 minutes for the dye to work in and at the same time be careful that not too much of the solution evaporates—otherwise there

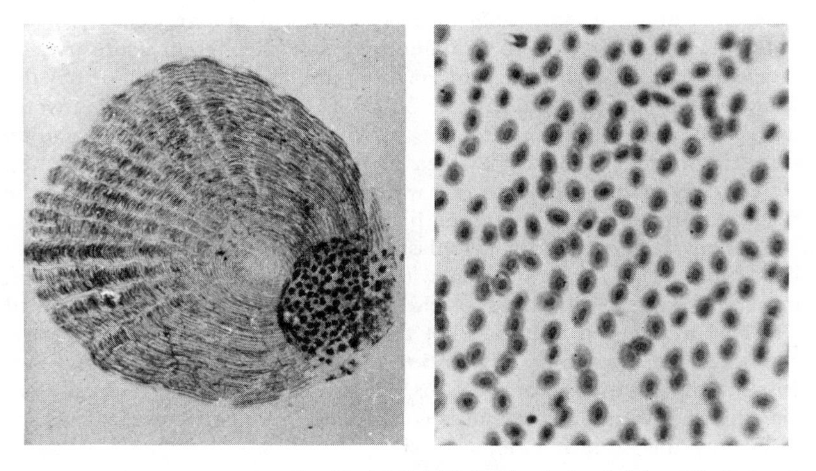

Fig. 91. (left) Fish scale. Fig. 92. (right) Fish blood, dyed. Magnified about 300 times.

will be an unsightly dye crust on the slide. It may become necessary to drip additional drops of the dye solution on it during the process.

After about 5 minutes drip—without pouring off the dye—boiled and cooled distilled water on the smear, about twice as much water as dye solution. Water and dye solution must be well mixed; this is not so simple as you cannot use a glass rod here without damaging the preparation. You must try to mix both fluids by tilting the slide back and forth.

Allow the diluted dye solution to work in for 5 minutes and then pour it off. The smear is then washed off with a squirt of distilled water (there must be no trace of the dye solution) and the slide is set up at an angle for drying. When the smear has dried fully, drop some resin on it and cover with the coverslip.

Even in the undyed smear, you can see that in a fish the red blood cells are oval, not round as in humans. The dyed preparation reveals another unusual thing: the red blood cells contain long, deeply-colored nuclei and are therefore real cells (Fig. 92). If you have an opportunity to observe the blood of different vertebrates and compare them, you will find that only in the mammals do the red blood cells have no nuclei.

In the same manner you can now examine your own blood. Note that the Wright's blood stain shows the white cells very clearly. You will also find that there are obviously several kinds of white blood cells: some have lobed, or "segmented" nuclei; others seem to consist of only one large, round cell-nucleus; a third have red-colored nuclei, etc.

For lesser demands, dyeing with hemalum-eosin is sufficient, as can be done with very many zoological and medical preparations:

Drop some Mayer's acid hemalum, which was filtered before use, on a dried blood smear. (Hemalum may be purchased in solution or powder form. In the powder form dissolve 5 gm. in 100 cc. of hot distilled water and filter after cooling.) After 10 to 12 minutes pour the dye solution from the smear and wash for 10 to 15 minutes in several changes of tap water; only after washing in tap water does the deep blue color of the nucleus show up in its full strength.

For contrast dyeing of the cytoplasm use a 0.1% eosin solution, prepared by diluting a 1% stock solution with distilled water. The eosin solution must work in for 3 minutes; then the surplus dye is washed out in distilled water, 3 minutes in 95% alcohol and 5 minutes in absolute isopropyl alcohol for dehydration and after 3 to 5 minutes in xylol. Mount in resin. Instead of eosin you can also use erythrosin. Erythrosin gives more glowing and deeper colors, but is considerably more expensive. It is used in exactly the same manner as eosin.

The frog is a very well suited animal for an introduction to the dissection of vertebrates, which forms the basis for your later microscopic, particularly histological examinations. If you are not able to get a frog, you can get a white mouse inexpensively in a pet store.

To kill a frog quickly and without pain, grasp it by the hind legs and slam its head against the edge of the table. Then sever the head with scissors and destroy the spine with a hot darning needle. If this method seems a bit bloody, the frog may also be killed with chloroform: place the animal in a not-too-large pressure jar, put a large, chloroform-soaked wad of cotton in with it and cover the jar with a glass plate. The frog becomes anaesthesized and does not waken if it remains in the chloroform-laden air long enough. Mice and other small mammals are also best killed with chloroform.

Fig. 93. Dissecting or wax pan.

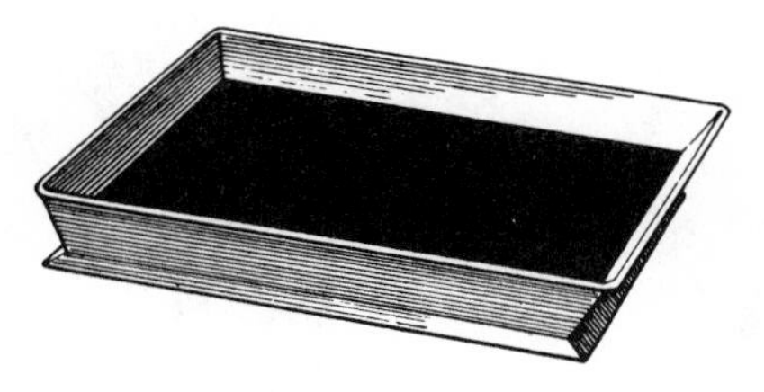

For dissection, the frog is placed in a dissecting pan with his underside up[14] and fastened down with strong, thick-headed pins. Stick the pins diagonally through the outstretched feet deep into the wax layer.

When dissecting you must remember these general rules:

1. When you dissect smaller animals, fill the dissecting pan with water, and perform the dissection under water; this makes dissection easier. As soon as the water gets dirty, replace it with clean water.

2. Damaging the organs with the knife or scissors must be carefully avoided; never cut away something which you do not recognize. Before an organ is removed, its position, attachment and connection with other organs must be exactly determined.

3. The instruments must always be kept clean and shiny. After use they are immediately cleaned and dried.

With tweezers, lift up the ventral skin a little and cut it from the crutch to the underjaw. With lateral cuts free the skin from the legs, pull it away from the flesh and pin it down alongside with needles stuck into the tray.

[14] Such a dissecting pan may be home-made: line a developing tray with a mixture of paraffin and beeswax. Used paraffin is not good; the addition of beeswax should be generous so that the stiffened mixture will not crumble.

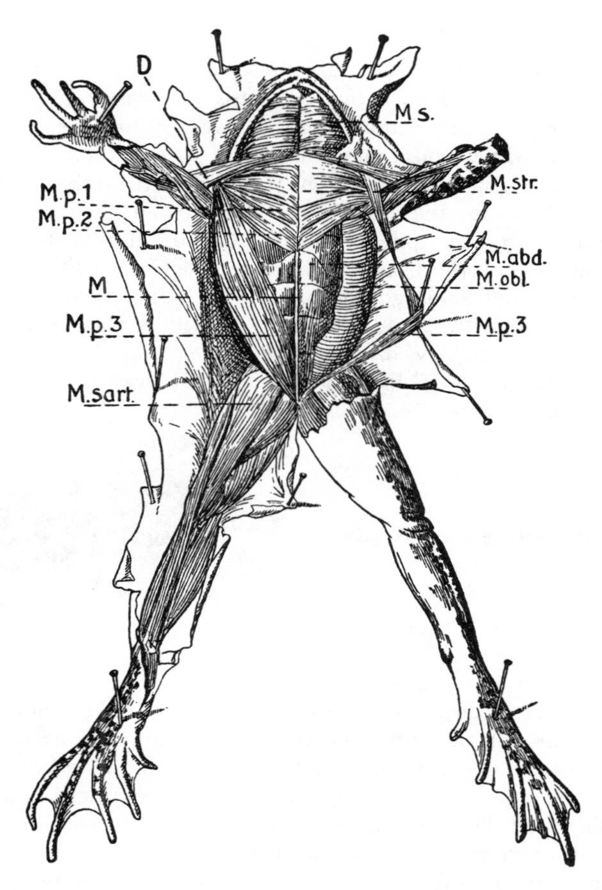

Fig. 94. Musculature of the green frog. D. Deltoid muscle. M.p.1. Fore part. M.p.2. Rear part. M.p.3. Ventral part of pectoral muscle. M.s. Lower jaw muscle. M.obl. Outer, stiff abdominal muscle. M.str. Sternoradial muscle. M.sart. Sartorial muscle. M.abd. Straight abdominal muscle. M. Median line.

This shows the musculature, which you see in Fig. 94. Now carefully pin aside the pectoral muscles and uncover the shoulder girdle and the breastbone. By cutting through the shoulder girdle on both sides near the upper arms, you can release it together with the breastbone. Some care must be taken here, so that the heart which lies underneath is not dam-

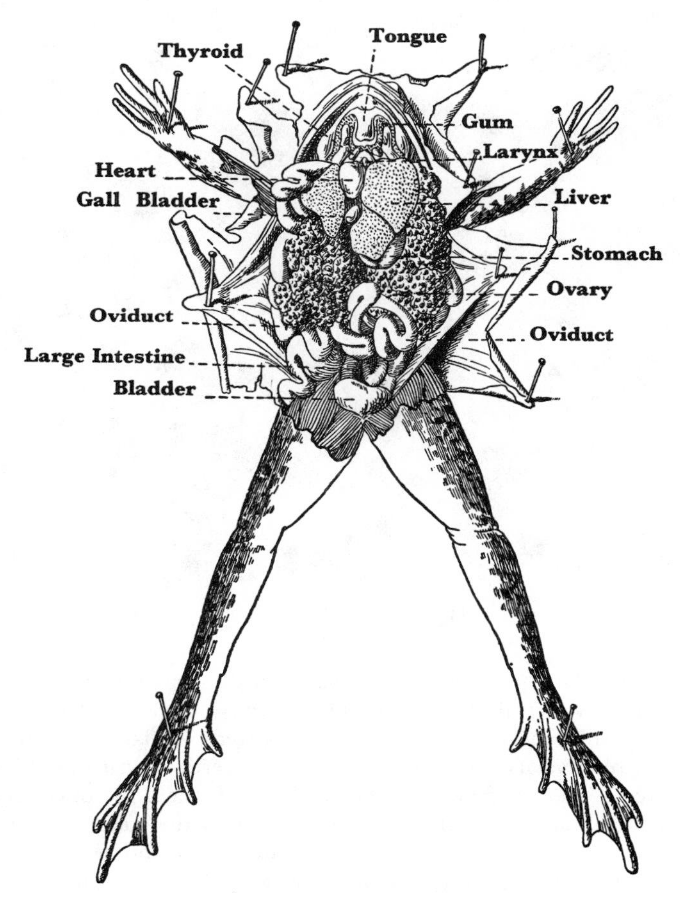

Fig. 95. Internal organs of a female grass frog.

aged. The fine membrane which covers the heart, the pericardium, is then carefully opened and removed. Now the entire heart lies free, and with a freshly-killed frog, it will still be beating. Cut it out carefully with its surrounding blood-vessels and place it in "frog physiological" (0.6%) common salt solution. In this it will keep beating for hours. After careful removal of the musculature in the neck region the tongue-bone may be seen as a delicate but large plate of cartilage. The thyroid glands lie on

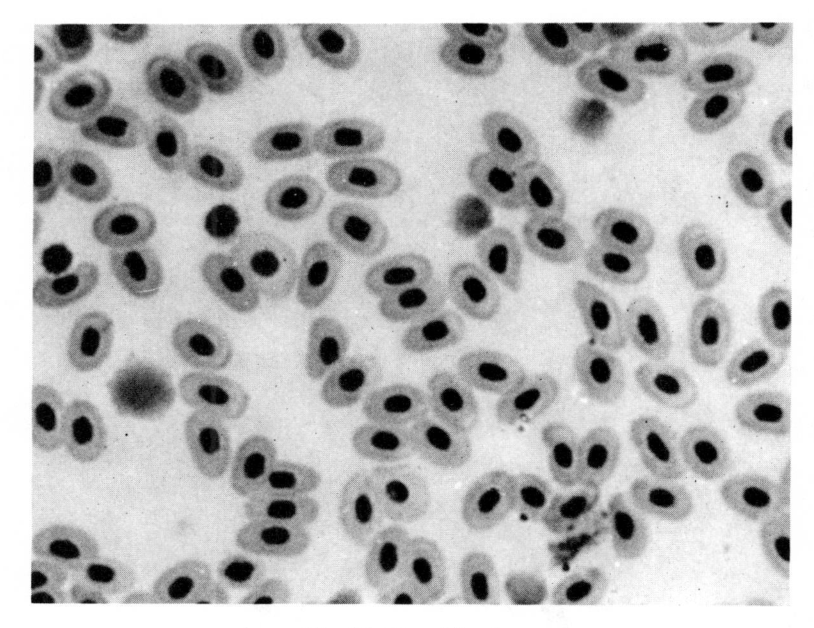

Fig. 96. Frog blood.

each side of the tongue-bone. Proceeding from the larynx you can easily find both lungs, which lie free in the body cavity. If you fold the large liver lobe inwards, you can see below the fair-sized bag-like stomach. Cut it open and observe its inner coating with the vertical wrinkles. If you fold the liver lobes over forwards, you see between them the dark green, rounded gallbladder. Under the intestines you find the spleen as a pink to reddish-brown body. Now observe the intestines and the different glands hanging from it in their natural positions. If you carefully squirt some water with a pipette into the cloaca, you can see the bladder which lies directly above fill up, making it easily recognizable by its form. Remove the intestine, all but the very end. The glands which are attached to the intestine come out with it and the separate parts of the intestinal tract are seen. Lastly, cut open the end of the intestine and examine its contents in water (see Fig. 100).

In a female, note under the liver on both sides of the middle line the greatly developed ovaries as dark-pigmented organs. In order to study them better, remove the organs which surround them. Once the ovaries are removed, the kidneys become visible: lying on both sides of the spine, elongated, flat reddish-brown organs. On the outer side of each kidney runs a whitish cord, the urinal tract. If you also remove the kidneys you can see beneath and at the sides of the spinal column the little yellowish-white calcium sacs. Spinal nerves are also to be seen, springing from the spine and extending to the rump and extremities. The heaviest one is the sciatic nerve, which passes through the upper thigh.

To expose the brain, remove the top of the skull carefully with a flat cut of the scissors. Note five parts of the brain—hindbrain (two sections), midbrain (two sections), and forebrain. If you carefully cut the nerves which leave the brain, you can remove it from the skull and place it in a watch-glass so that you may also study the underside of the brain.

Such a dissection is made to learn about the different body parts. Naturally you also get a great deal of material in this way for microscopic preparations.

FROG BLOOD

The blood which flows out when the head is severed provides your first preparation. Make a blood smear and treat it as for fish blood. Upon examination you immediately see the large, oval, nucleus-bearing red blood cells.

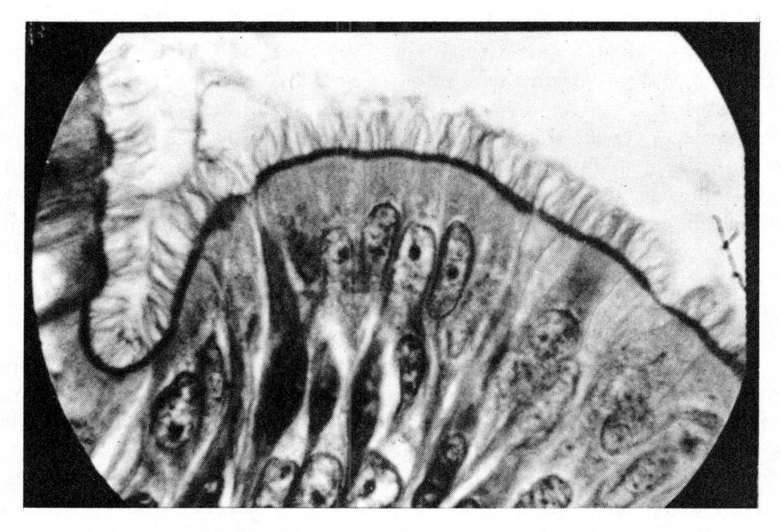

Fig. 97. Ciliated epithelium from the mucous membrane of a frog.

CILIATED EPITHELIUM OF THE FROG'S MOUTH

Remove a freshly-killed frog's lower jaw and cut out a little of the mucous membrane from the inside of its mouth. Transfer this quickly to a slide in a drop of 0.6% salt solution and cover with a coverslip (for frogs, the correct physiological salt solution is 0.6%, and not 0.9%, as for mammals). If you focus on the edge of the preparation, you can see with high magnification the fluttering cilia waving back and forth, like a field of oats in the wind. Since you cannot distinguish the epithelial cells on which these cilia rest, because of their rapid movements, slowly kill the tissues by adding a drop of formalin. The lively motion ceases and the individual cilia stand out clearly. To make a permanent preparation, fix the portion of mucous membrane in formalin 1:4 (1 part of formalin, 4 parts of tap water) for several hours. Then transfer it to several changes of tap water and afterwards place in distilled water. Dyeing: hemalum 10 minutes; wash out in tap water 15 to 20 minutes. Contrast dye in 0.1% eosin solution 3 to 5 minutes, quickly wash in distilled water, dehydrate in ascending alcohol stages or without the stages in methyl glycol, terpineol or methyl benzoate for 10 to 20 minutes, rinse with xylol, then mount in resin.

STRIATED MUSCLES

Take off a little bit of the skinned upper thigh of the frog and put it into a drop of physiological salt solution. With two preparation needles, split the bundle lengthwise into two equal halves. This is easily done if you work the needles in opposite directions along the fibers. With the other piece, repeat this process and continue until you have a number of fine fibers. Now put on the coverslip and view the preparation with weak magnification at first, then stronger. Note the fine striations of the muscle fibers, which are caused by very fine threads, the muscle fibrils. The muscle fibrils pass lengthwise through the fibers, and each fibril is composed of alternate light and dark bands. Because each band of the fibril lies inside the fiber at equal height, the entire fiber seems to be striated.

If you examine the muscle fibers in tap water you soon see how in some cases the sarcolemma, the delicate structureless membrane which surrounds each muscle fiber, lifts off bubble-like as a delicate skin. To make a preparation, fix a tiny piece of the muscle in formalin 1:4, wash it in tap water, transfer to distilled water, dye with hemalum, blue in tap water, dehydrate with alcohol or methyl glycol and finally transfer to terpineol. Then place a drop of terpineol on a slide and put the bit of

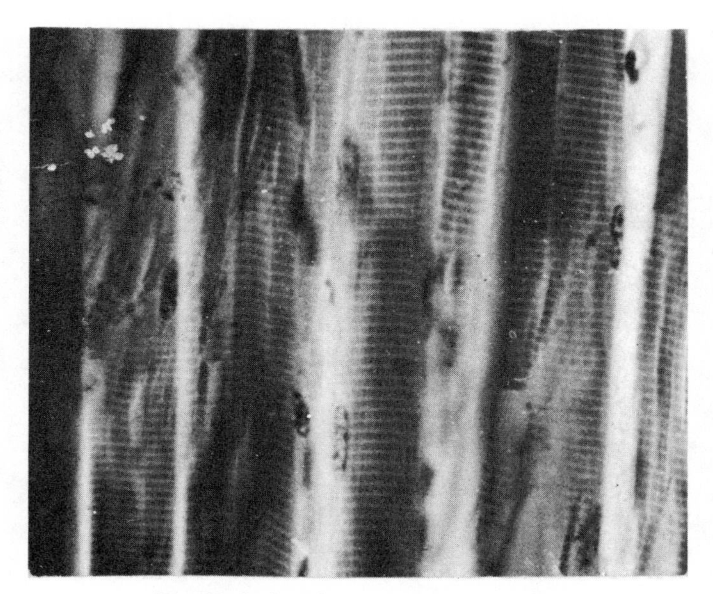

Fig. 98. Striated musculature of a frog.

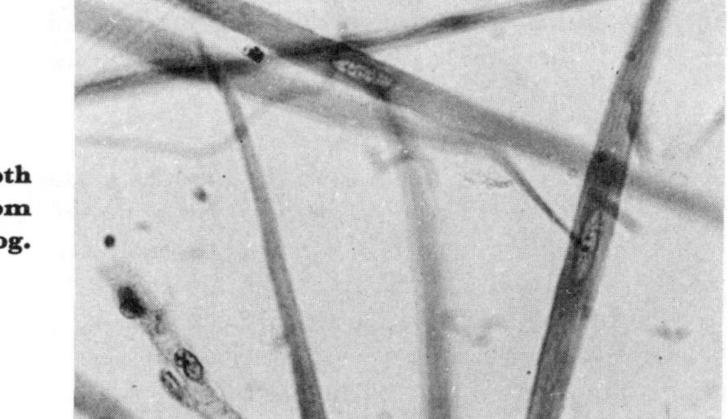

Fig. 99. Smooth muscle fibers from the bladder of a frog.

muscle which is already drenched with terpineol in it. With two preparation needles, pick apart the bit of muscle as before. If enough fibers have been separated, the coarser bits are taken out, the surplus terpineol absorbed with filter paper and resin dropped on it. After putting on the

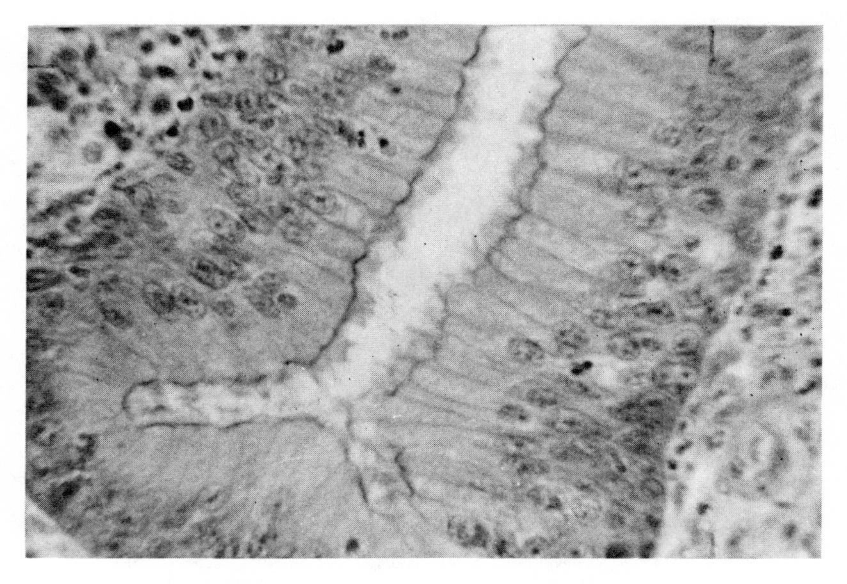

Fig. 100. Columnar epithelium intestine of a frog.

cover-glass the permanent preparation is finished and you can easily study the muscle fibers with cell nuclei and sarcolemma. In many cases the ends of the muscle fibers are split into fine fibrils.

SMOOTH MUSCLE FIBERS

Get smooth muscle fibers from the bladder of the frog. Pull apart a bit of the bladder wall in physiological salt solution. Because the spindle-shaped, elongated muscle cells cling to each other with extraordinary tenacity, you have to pick apart the material into a fine mass of tissues in order to isolate the single cells. Each smooth muscle fiber has a cell nucleus, which shows up very plainly when acetic acid is added. A permanent preparation can be finished as for striated muscles.

COLUMNAR EPITHELIAL CELLS

Beautiful epithelial cells are found in the mucous membrane of the small intestine. Cut out a piece of the small intestinal wall and pull away the mucous membrane with a sharp scalpel or with the razor. Rinse the bit of membrane in physiological salt solution and store for 12 to 24 hours in a closed container with 30% alcohol. Pick the tissue apart on the slide in 30% alcohol and continue the procedure as for ciliated epithelium or striated muscle.

NERVE FIBERS

Prepare these by pulling out the sciatic nerve from the upper thigh, cutting off a small piece and placing it on a slide in a drop of physiological salt solution. With two needles pick the nerve apart lengthwise. Under the microscope note the medullated nerve fibers with axis-cylinders, myelin sheaths and the neurofibrils.

To make a permanent preparation, place a bit of the fresh nerve on the slide without any additional fluid and spread it out as quickly as possible with two preparation needles into a fine white film. On this put 1 or 2 drops of osmic acid solution, allowing it to work in for 15 to 20 minutes. Then wash off the osmic acid with repeated drops of distilled water and mount in glycerin jelly. Unfortunately osmic acid is so expensive that it is out of the question for most microscopists. When working with osmic acid you must be careful because the vapor is very irritating to your mucous membranes.

INTESTINAL INFUSORIA OF A FROG

Take some of the intestinal contents from the frog with as much intestinal mucus as possible and place it in physiological salt solution on a slide. You will find large, oval, flattened ciliate micro-organisms with many nuclei and a distinct longitudinal striping (*Opalina ranarum*, Fig. 102). On another slide spread out some intestinal contents without any added substances and handle it as the earlier instructions for Infusoria.

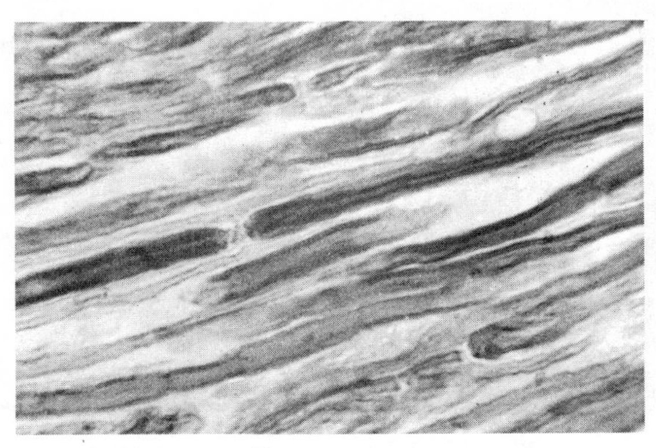

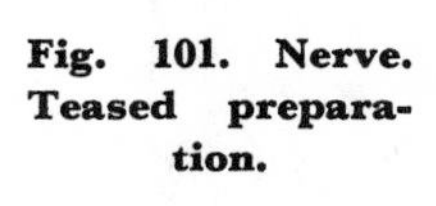

Fig. 101. Nerve. Teased preparation.

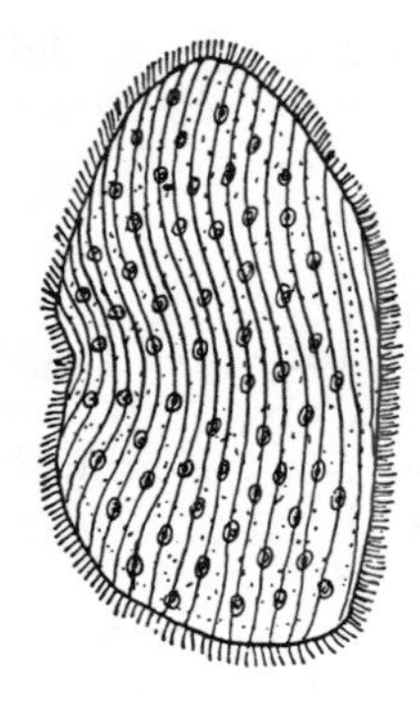

Fig. 102. *Opalina ranarium* **from the rectum of a grass frog.**

CARTILAGE

The butcher can provide a rib end of a calf, about 3 cm. long. Make thin slices with the razor exactly as you have learned to do for plant sections. In order not to tear the tissues, moisten the cut surface and the razor blade with physiological salt solution. The sections are placed in a watch-glass with physiological salt solution. At least one slice should contain some gristle surface.

Transfer the thinnest sections to a formalin and water solution (1:4),

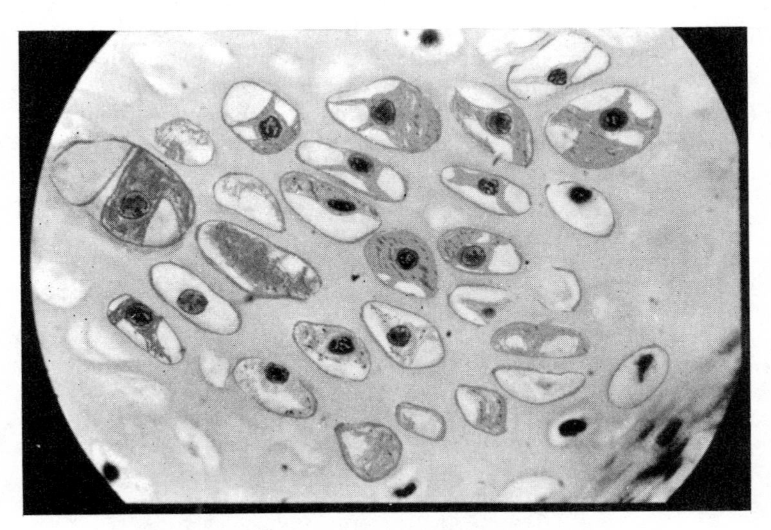

Fig. 103. Gristle. Drawn from a microphotograph.

wash them off after half an hour with tap water and dye with hemalum. Further steps are as for ciliated epithelium.

In your examination look for the membranous covering (perichondrium) or gristle skin and the cells which lie in the hollows in the matrix of the gristle.

BONE

Trim off some slivers 1 to 2 mm. thick from the edge of a narrow bone across the grain with a hacksaw. Rub these slivers on a well-moistened piece of sandstone in a circular motion until the under surface is ground even. Then wash, wipe, and polish it. To do this, stretch stiffly a piece of chamois, sprinkle it with powdered chalk and rub the bone fragments on it in a circular motion. Finally wipe off the polished side with clean chamois and cement it on a slide. Use Canada balsam as a cement, buying it in solid pieces (do not buy it mixed); this must be dried carefully over a flame. Melt the Canada balsam on the slide and press down the polished side into the softened resin. The balsam must come around the bone on all sides; this can easily be accomplished with repeated heating and pressing down with your finger. Once the Canada balsam has become hard, grind down the upper surface of the bone on the stone until it has

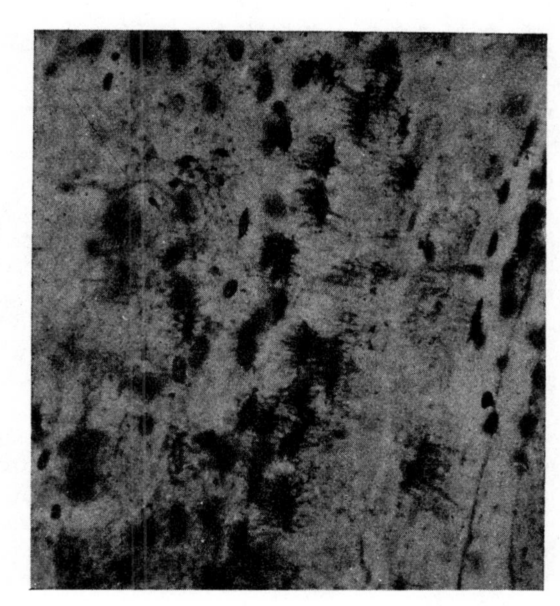

Fig. 104. Bone sliver.

become paper-thin. Polish the upper side in a similar manner and mount the sliver in glycerin.

If the sliver is thin enough, you can see under the microscope the cross-sections of the Haversian canals, which are enclosed in the bone lamellae (in cross-section they look like concentric rings). These spaces, in which the bone cells lie, are filled with air and give a sharp contrast. These "bone canals" are connected to each other with branched offshoots.

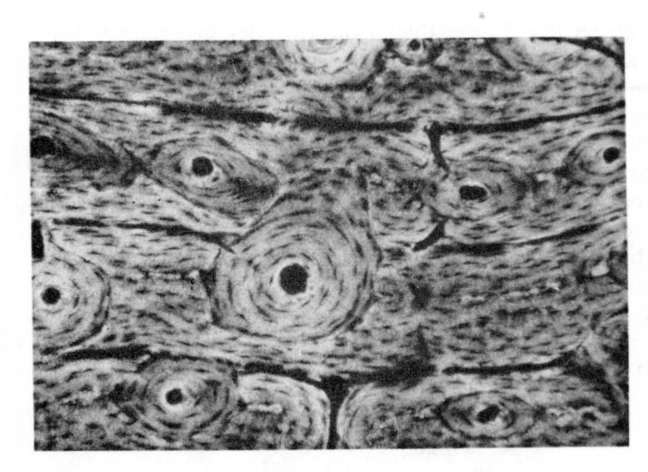

Fig. 106. Bone section.

BONE SECTION

Saw a hollow bone of a small mammal into little pieces. It is important that the bone be fresh. In order to cut the bone, you must decalcify it. For this, use trichloroacetic acid which fixes it at the same time and saves a step in the operations. Put the bone pieces in a 5 to 10% solution of trichloroacetic acid in water to which you have added 10 cc. of formalin to each 100 cc. of water. The bits of bone are repeatedly shaken in the acid and every few days the used solution should be replaced with fresh solution. The pure trichloroacetic acid from which you prepare the solution must always be well sealed, as it absorbs water from the air and becomes diluted.

For small bones you can try to cut them in about a week. If the bones are quite soft, the trichloroacetic acid is washed out of the bone tissues in several changes of 90% alcohol (at least 24 hours longer does no harm).

Dye the slices with hemalum or still better in strongly diluted Delafield's hematoxylin. Dilute 1 cc. of the purchased hematoxylin solution with 50 cc. and leave it in this for about 5 hours. The dye is then washed out with tap water. Dehydrate and mount in synthetic resin in the usual manner.

PORK LIVER

Cut strips about 1 to 2 cm. long and ½ cm. wide from a fresh liver and fix them in formalin 1:4 (1 part of 40% formalin, 4 parts of tap water). After 24 hours, place the pieces in 50% alcohol, then in 70, 80 and 95% alcohol (each stage at least a day). When the liver pieces have hardened after several days in 95% alcohol, make sections as thin as possible with the razor. It is sometimes advisable to moisten the cutting surface and blade with 70% alcohol.

The sections are next placed in a dish with 70% alcohol, then after a 50% alcohol bath into distilled water, dyeing with hemalum-eosin.

The finished preparation shows red-dyed connective tissues which divide the section into many irregular, many-cornered zones. Every zone is the cross-section of a liver lobe. In the middle of each zone there is a blood vessel (the main vein). In the connective tissue note cut-open bile ducts.

You can satisfactorily process most animal organs. It is a universal misconception that animal tissues can only be cut with the help of a microtome. The experienced microscopist can make perfect sections with well prepared animal organs. The only thing is that it requires a little more practice than for plant materials.

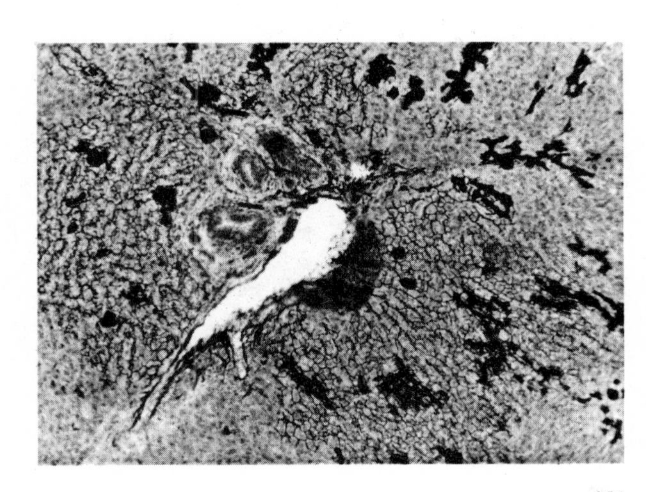

Fig. 107. Pork liver.

Often it is advisable when working with animal tissues to soak them in terpineol as well after hardening them in alcohol. The pieces then become transparent and acquire a very good consistency for cutting. Also, you do not need to be afraid that they will dry out while cutting because the oily terpineol evaporates very slowly. The sections are then placed first in 95% alcohol and then through descending alcohol stages into distilled water.

LIVER CLAMPS

Because liver which has been hardened in alcohol cuts very well, you can use it in place of elderberry pith for clamping small objects. Harden pieces 2 to 3 cm. long and 1 to 2 cm. wide a few days in spirit, which is changed after the first day. To use, split about half the length of the liver strip, clamp the object in it and cut along with the liver. In alcohol the bits of liver are easily removed from the sections.

8. Bacteria

People are apt to connect the word "bacteria" with the organisms carrying plague, cholera and other diseases, and perhaps with decay and decomposition. However, the disease organisms are only a very small group in the huge horde of bacteria and by far the least important. More important than the parasites are the inhabitants of decaying matter and the causers of decay, the bacteria in water, those that live in the intestines and those that live in a symbiotic relationship with other organisms.

The microscopic examination of bacteria is intriguing, especially if combined with the biological examinations of culture procedures. Every microscopist should know the most common bacterial forms and should have practised at least the most important examination procedures once. Unfortunately there is a lack of space to describe the culture as well as the examinations.

Where do you find bacteria? Bacteria occur in practically all parts of the globe. Bacteria can be found in the water, the earth, food, even in hot springs and petroleum. For your examinations, you will want to draw upon sources where you will surely find bacteria in large quantities: decaying substances are always swarming with bacteria of many sorts.

The most easily obtained source of bacteria is the surface coating of a hay infusion. Other bacteria can be found in decaying manure or in a mixture of garden soil, water and a few pieces of meat, left to decay for a few days or weeks in a warm spot.

Take out a drop of the material to be examined, place it on a slide and view it with a high-power objective. Note a wild swarm of tiny rods, balls and spirals—the bacteria.

Some bacteria are capable of movement—they are provided with tiny hairs. You cannot, however, observe the hairs themselves, but can see the motion of the bacterial cells. (To show the hairs there are special, none too easy, methods.) Note how single bacteria move smoothly in one direction across the field of vision. This directed motion is active—not like Brownian movement, a trembling dance in one spot which is performed by the non-ciliate and therefore non-motile bacteria.

The globular bacteria are cocci; the spiral ones are known as spirillae; the rod-shaped ones are bacilli. The rod forms or bacilli often produce resistant bodies called spores.

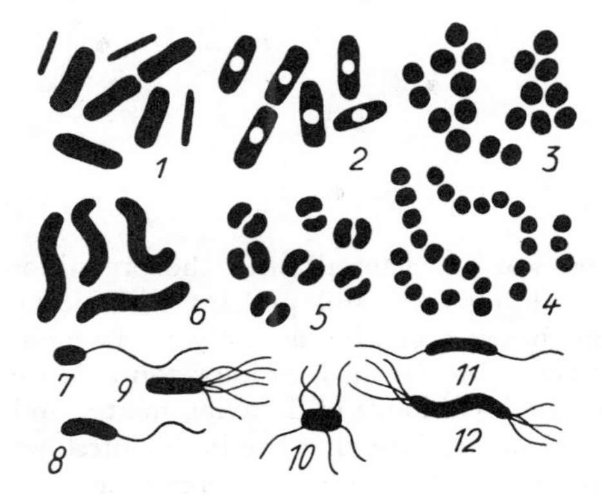

Fig. 108. Several forms of bacteria. 1. Bacteria without spores. 2. Bacilli with spores. 3. Staphylococci clusters. 4. Chains of streptococci. 5. Diplococci. 6. Spirillae. 7-12. Flagellates.

DYEING BACTERIA

Take a well-cleaned cover-glass and put a drop of bacteria-laden substance on it, then spread the drop over the entire surface of the coverslip. The smears may be dried in the air (bacterial preparations are among the few objects which can dry out without any damage).

Before dyeing you must fix and degrease the smear. For fixing, grasp the cover-glass with tweezers and draw it—smear on top—three times through a small flame. The cooled cover-glass is then degreased by dipping it several times in ether and then allowing it to dry again. (Degreasing is not absolutely essential, but very desirable; smears which have not been degreased sometimes show ugly, difficult-to-remove dye spots.)

For dyeing use Löffler's methylene blue and carbol fuchsin. Methylene blue dyes the cocci particularly well; carbol fuchsin dyes the spirillae and bacilli. There is little sense, however, in using both dyes on the same smear if there is no particular purpose in so doing. It is better to dye one with methylene blue, and the other with carbol fuchsin.

Drop the dye solution generously on the smear and allow it to work in for five minutes. Then pour off the dye and rinse well with distilled water. Allow the smear to dry in the air, then put some resin on a clean slide and place the coverslip face down on to it.

The microscopic examination will show some bacteria to be dyed a sharp blue, some red.

USE OF THE OIL-IMMERSION OBJECTIVE

The examination of dyed bacteria gives you an excellent opportunity to get acquainted with the use of an oil-immersion objective (1/12 in.). In contrast with the usual dry objective the space between the cover-glass and objective is filled in with a fluid. In this way the high magnifying power of this special objective is utilized. The highest magnifications and —what is more important—the best possible clarity can be attained with oil-immersion objectives.

If you know how to work with an oil-immersion you can examine the finest structures which it is possible for the microscope to show. As an immersion oil, use a good synthetic immersion oil made by a reputable firm.

Such objectives are generally used with an immersion oil of known refractive index. A good oil-immersion lens has a numerical aperture of 1.3; that is to say that with it, magnifications up to 1,300 diameters may be used without going over the limits of useful magnification. To use higher magnifications is, however, in most cases useless, even with an oil-immersion lens (see page 15).

Many beginners make the mistake of using the oil-immersion objective without immersion oil. This results in a poorer image than is obtained with a dry objective.

High-power objectives, such as oil-immersion objectives, give very dim images. You should not work, therefore, with daylight, but with an artificial source of light.

To use the oil-immersion, first focus a spot in the preparation with one of the weaker dry objectives. The tube is then screwed high up, and on the coverslip place a drop of immersion oil with a glass rod. Now lower the tube with the oil-immersion objective very slowly, watching the objective from the side. Look into the microscope only after the lens has made contact with the oil and focus the picture sharply. If possible the focusing should be done with the micrometer adjustment only, as contact with the coverslip must be avoided. The preparation should be shifted only slightly, once the oil-immersion has been adjusted, to avoid smearing the coverslip with too much oil.

The focal length of the immersion objective is very short. If too thick coverslips are used, it may happen that the coverslip is thicker than the focal length. Such a preparation cannot be examined with an oil-immersion objective. You must therefore make a rule that all preparations which are to be examined with an oil-immersion objective must be covered with selected thin coverslips. Unfortunately the coverslips which are bought are very variable in thickness. You can find coverslips of many different thicknesses in the same box, and you must select the thinnest of these. Some makers sell them already graded and can supply you with coverslips of guaranteed thinness.

When the examination is finished, the tube is screwed high and the oil adhering to the front lens is wiped off with a thoroughly clean linen cloth. Then take an unused part of the cloth and dampen it with a drop of benzine, and with this remove the rest of the oil from the lens. Lastly wipe off everything with a dry cloth and check to see if the front lens is really clean. A few remaining traces of oil are wiped up with a cloth that has been dampened with benzine. When cleaning there must be no pressure on the lens. Xylol must never be used under any circumstances as it damages the lens cement.

The coverslip which was smeared with oil can be cleaned with xylol. However, here too it is better to use benzine.

The immersion oil is best stored in a specially constructed, tightly sealed immersion oil bottle.

NITROGEN-FIXING BACTERIA

If you remove a sweet pea or any other bean-like plant out of the soil, even a rough examination will show many round, egg-shaped or branched little nodules, varying from the size of a pinhead to the size of a pea, depending on their age. Cut one of these nodules in half with a scalpel, scrape some tissues from the cut surface into a drop of water, stir well and spread thinly on a clean coverslip. The air-dried smear is dyed with Löffler's methylene blue.

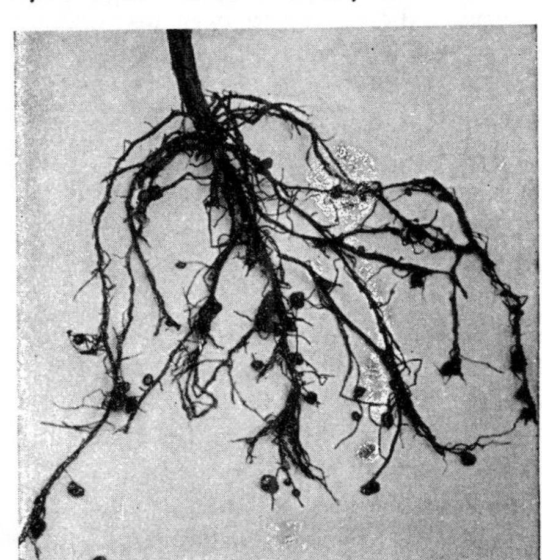

Fig. 109. Root of bean plant with root nodules.

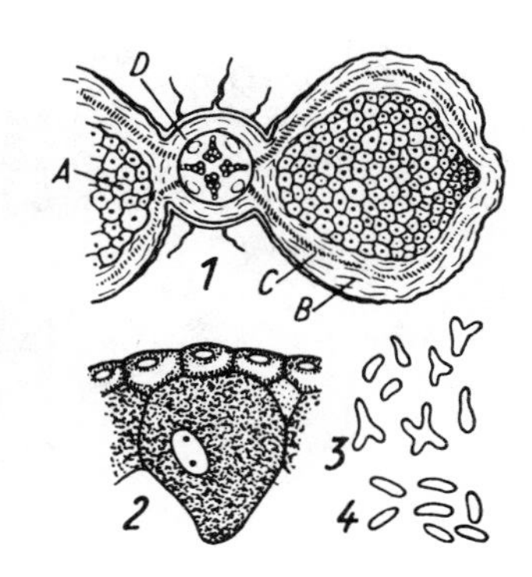

Fig. 110. Bacterial nodes from a root of *Vicia faba* (soybean). 1. Young nodes. A. Large-celled tissue with tissues filled with bacteria. B. Node rind with canals. C.D. Root. 2. A node cell filled with thousands of bacteria with neighboring uninfected cells (above). 3. Forms of "bacterioids" and 4. Unchanged bacteria (*Rhizobium radicicola.*)

Note the irregularly-shaped structures (Fig. 110)—bacteria, which live in the root nodules. These bacteria—in this case *Rhizobium radicicola*—live in a symbiotic relationship with the plant. They make possible the extraction of nitrogen from the air and bring a part of the nitrogen into the host plant, which derives a great deal of benefit from this bacterium. It is therefore possible to cultivate sweet peas in a soil which is very poor in nitrogen. Lately it has been attempted to inoculate soil with pure cultures of these bacteria and promote such symbiosis.

In very young nodules you will find the bacteria as slender little rods. Only in the older stages do they change their shape and become larger. Finally the host plants digest these bacteria. Nodules which have been preserved in alcohol may also be used.

SULPHUR BACTERIA

In drainage canals and muddy waterholes you can see even with the naked eye white, veil-like coatings. The waters often have a sulphurous odor, and the coatings are in many cases sulphur bacteria.

At the spot where they are found, fill a preserve jar one-third full with the foul mud, then the remainder with water from the same spot and transfer some fragments of the white skin to the glass. At home store the container in a not too bright spot.

Place a drop of water from the storage jar on a slide; take some of the white substance and put it in the drop and put a cover-glass on it.

Using a low magnification, you can see all sorts of micro-organisms swarming in the foul water: many Infusoria, threadworms and algae. In between there are colorless long threads, which contain more or less numerous, black-ringed balls. The balls consist of sulphur. A higher

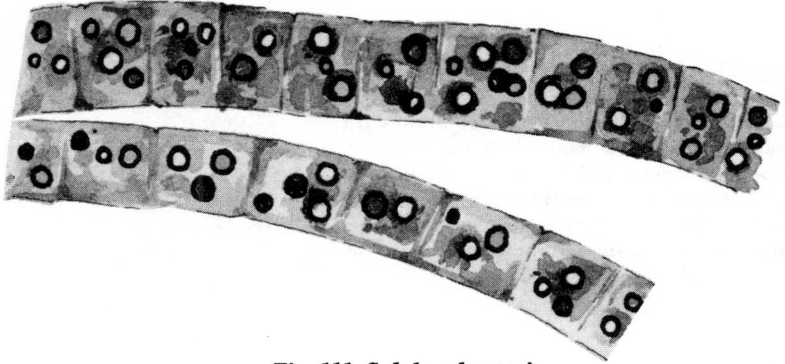

Fig. 111. Sulphur bacteria.

magnification furnishes more exact details of the construction of these thread bacteria (Fig. 111).

To make a permanent preparation, allow a richly populated drop to evaporate. After fixing with heat, dye with carbol fuchsin.

DYEING BACILLI SPORES

If you examine the coating on an older hay infusion, you will observe many moving slender rods. They are mostly *Bacillus subtilis*, the common bacillus. With the help of an oil-immersion lens you can see some cells with rounded, very refractive little bodies in this living preparation: spores, the persistent form of the bacilli. These spores are developed when conditions are not proper (lack of nutrition, lack of oxygen, etc.) inside the cells; this process uses up much cytoplasm. When the cells fall apart, the spores are freed. Because the spores are very durable (they can dry out and resist many other conditions, even boiling heat), the bacilli in this form can tide over unfavorable conditions. The familiar rods grow again from these spores.

You can find very good spores in butyric acid bacilli, which can be obtained by very simple culture:

An unwashed potato, which if possible still has some dirt clinging to it, is stabbed deeply with a knife in several places. The potato is put into a jam jar in water, so that the water covers it. The butyric acid bacillus in the potato, which you want to culture, belongs to the so-called anaerobic bacteria, which do not tolerate oxygen. The water over the potato serves to keep out the oxygen from the air. After one to three weeks you find inside the potato a stinking, creamy mass around which gas bubbles form which drive the potato to the surface. If you remove a little of the soft mess and mix it in a drop of water, you will then probably discover a great many relatively large spores.

To dye the bacilli spores you cannot make use of the usual bacterial dyes because the spore membrane lets through the dye only after special preparatory measures are taken. Use the following procedure.

The dried coverslip smear is well fixed by drawing it through a flame several times. The coverslip is then placed in a 5% chrome acetic fluid solution. The chrome acetic fluid is shortly washed off with water and the entire smear covered with carbol fuchsin. Then with a pair of tweezers take the coverslip and hold it over a small flame. As soon as vapors rise from the carbol fuchsin, remove the coverslip from the proximity of the flame and hold it for a minute in such a manner that the solution remains hot, but does not come to a boil. The dye solution is then rinsed with water, the coverslip immersed for a few seconds in 5% sulphuric acid and immediately washed off with water again. The sulphuric acid process is the most important one; the sulphuric acid bleaches the bacteria once more. As the spores hold the dye more tenaciously than the bacteria, you must catch the period of time when the spores are still dyed and interrupt the bleaching process by washing off with water. If the smears remain in the sulphuric acid a little too long, the spores will also be bleached. After the sulphuric acid has been washed off, dye a few minutes with Löffler's methylene blue in the usual way, rinse off with water and allow to dry in the air. The dried smears are then mounted in resin. Successful preparations show the spores red and the bacterial bodies blue.

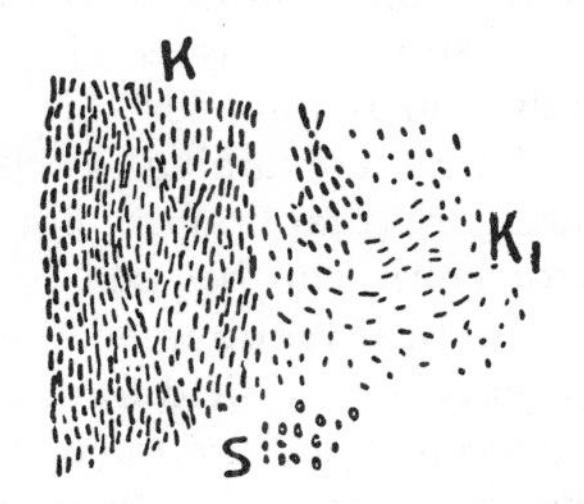

Fig. 112. Bacteria from the surface film of a hay infusion (*Bacillus subtilis*) K. Undisturbed and K_1 disturbed film. S. Some rods and spores of the hay bacillus. Magnified about 600 times.

9. Microphotography

If you want to make good microphotographs you must be fully acquainted with the basic optical laws of the microscope as well as those of the camera. Above all, microphotography requires a good knowledge of lighting technique. One seldom comes across really perfect microphotographs, and for scientific purposes a drawing is in most cases preferable. It will provide much joy nevertheless to many microscopists to make a photographic record of unusual preparations and objects which were prepared with great pains.

The following brief directions can serve only as the simplest of introductions. If you want to go into microphotography thoroughly, refer to special literature on the subject.

LIGHTING

To obtain the best microphotographs it is essential to use what is known as "Köhler illumination". This is also ideal for exact visual observation. Unfortunately you require for this a rather expensive low-voltage "research-type" lamp, containing a clear filament bulb, an auxiliary lens and a large iris diaphragm. But any very bright light source, such as a projector lamp, can be used provided the apparatus is arranged to satisfy certain optical requirements.

In order to attain an evenly-lighted field from an ordinary lamp, place a piece of frosted glass between it and the microscope. When the lighting is focused, the frosted glass is removed. Daylight is out of the question for microphotography: even the small microscope lamps with direct plug-in connections are usually too weak.

APPARATUS FOR MICROPHOTOGRAPHY

The quality of microphotographs is dependent to a higher degree on correct lighting than on the camera used. It is possible to take outstanding pictures with the most primitive equipment because the auxiliary parts of the camera are not used and in many cases even the lenses are removed.

There are microadapters for most small cameras, with the help of which the microscope and camera can be connected. The directions of the various manufacturers give instructions for the proper procedures. For all microcameras that are not firmly attached to the microscope, vibration must be avoided during exposures.

LIGHT FILTERS AND LIGHTING

Only in the rarest instances can you work without a light filter when photographing. Make this a rule: undyed preparations get photographed with a green filter, and dyed ones with a green or yellow filter. (A yellow filter is used only when the green filter delivers a light which is too hard.)

The question of exposure time is a very difficult one to answer. The exposure meter is not much help in microphotography; in the beginning there is much trial and error. If you use plates for photography you can easily determine your trial lightings if your camera has a focal-plane shutter, for a series of exposures can be made on the same plate. With the first exposure, open the shutter fully. Before the second close it slightly, so that a strip of the plate is not exposed. Then with the third cover another strip of the plate, and so on. In this way you get a picture which in its stages shows different exposure times. If the same exposure time has been chosen for each exposure (for example, 3 seconds), you can easily determine at which point the structures are clearest; then use that exposure. If for technical reasons, you cannot make use of this method, then three or four test exposures must be made with different exposure times. Here you must remember that exposure times in microphotography are usually very long. Depending upon the light source, filter, object and magnification you will have to choose exposures of between 3 and 50 seconds. Once you become familiar with your apparatus, you can then judge the exposure time better; the experienced operator will seldom make mistakes in his exposure times.

Do not use highly sensitive films and plates which are used for ordinary photography for your negative material. Because the exposure times are always long it is better to use a finer-grained negative material of light and medium sensitivity. Unfortunately, microphotographs show very little depth of focus, for the camera does not accommodate itself like the human eye and it is obviously useless to fiddle with the fine adjustment during an exposure. The thinner the object, therefore, the sharper the photograph that can be obtained.

Color microphotography requires a great deal of experience. For color photographs you need a particularly good objective: the so-called apochromatic lens, which requires the use of special eyepieces (compensating eyepieces). We can, however, only mention this special subject here, and must forbear also to go into more technical matters, such as electronic flash photography. More advanced books should be consulted for information on such subjects.

Appendix

Microtome Technique

In order to furnish extremely thin, absolutely even sections, you require a special apparatus, the microtome. With the help of this "slicing machine" you can make slices without holes of such a thinness as you could never get by hand cuttings, even with much practice. Naturally you can see much more in a series of sections than in single cells; in this way it is possible, for instance, to reconstruct the entire inner structure of an organ by means of a series of sections.

Microtomes are precision instruments and are therefore expensive; nevertheless for many purposes a simple hand microtome is sufficient, as described on page 99. It can however never be too strongly impressed that every microscopist should learn the manual cutting technique thoroughly before he attempts to make microtome sections. The feeling for every cutting operation can best be obtained by becoming practised in the use of the hand razor. Besides, in many cases, particularly with objects from the plant world, hand cuttings are sufficient even for scientific purposes.

If you wish to section any object on the microtome, you must first prepare it in a definite manner. Most of the materials—whether they be of animal or plant origin—are much too soft to be cut without any further preparation. You must therefore—after previous painstaking fixing—embed them in paraffin or celloidin, if you do not prefer to give them sufficient hardness by freezing. First you must acquaint yourself with the preparation of objects for microtome operation.

Plant and animal structures which have been removed from their natural surroundings change so quickly that in a short time they are no longer suitable for microscopic examination. This process, known as autolysis, you must try to slow down. As soon as they are removed from the organism, place the tissues to be examined in a fluid which kills the cells and at the same time preserves their structures as naturally as possible. From the multitude of known fixing solutions, select a few which have proved themselves for decades; almost all will give good results to start with.

When fixing—as with all the other subsequent tasks—you must work with great caution. Sloppy work only leads to failures!

In order that the fixing fluid will penetrate the object equally from all sides, place it in a small glass vessel filled with the solution covering a little glass-wool. The volume of the fixing solution should be at least 50 times that of the object to be fixed.

The time required for fixing varies greatly, depending on the nature of the fixed object. You must be careful that the bit of organ is fixed all the way through without being left too long in the fluid. Larger objects are much more difficult to fix and later are more tedious to section than small ones. If possible, it is therefore best to choose pieces of from $\frac{1}{2}$ cm. to 1 cm. in thickness.

The most easy-to-use fixing solution is formalin, the 40% solution of formaldehyde in water. The commercially sold formalin is thinned with tap water at a proportion of 1:4 and the objects are left in it from one to several days, depending on size. After this they are washed in several changes of 50% alcohol or in water for 2 to 3 days.

Excellent results can be achieved with Bouin's fixing mixture. To prepare it, mix together, just before using, 15 cc. of saturated aqueous picric acid solution, 5 cc. formalin and 1 cc. of acetic acid and fix smaller objects in this for 2 hours, larger ones 1 to 2 days. After fixing, the objects are washed for 1 to 2 days in 70% alcohol which is changed 3 times.

For examining the cytoplasmic structures, a very good substance is potassium bichromate-acetic acid. Shortly before using, mix 100 cc. of potassium bichromate solution with 5 cc. of acetic acid and fix in this for 24 to 48 hours. In this case, wash with tap water, at best in running water for at least 12 hours. If the constantly splashing sound of running water is disturbing, help can be given as shown in Fig. 113: a cork with a hole drilled in it is placed in the tap and a thread runs from the cork to the surface of the water. The water runs along the string with scarcely a sound if the tap is not turned on too far.

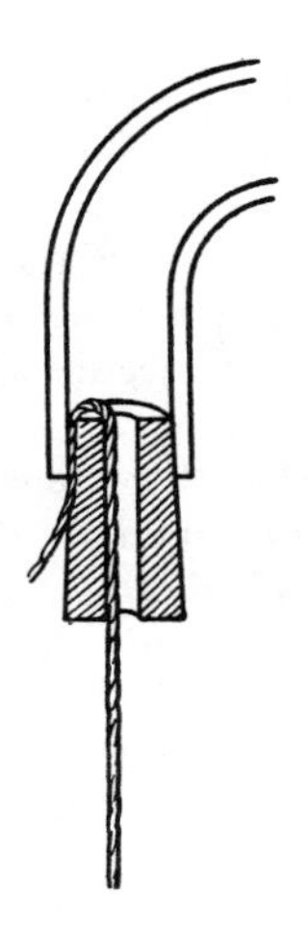

Fig. 113.

After fixing the organic materials in potassium bichromate solution, and until you embed them in paraffin or celloidin, they must be kept in the dark (as in a cabinet) because light causes bothersome precipitations.

There are various sublimate mixtures among the most commonly used fixing materials. Because corrosive sublimate (chloride of mercury) is a dangerous poison, its use requires great caution. Very good results follow the proper use of acetic acid sublimate, a mixture which is indicated especially in embryological examinations. Mix 100 cc. of saturated sublimate solution with 5 cc. of acetic acid to fix very small objects for half an hour, larger ones up to 24 hours.

The subsequent treatment of objects fixed in sublimates is somewhat complicated. Wash in 70% alcohol and transfer to 80% alcohol to which enough iodine in aqueous potassium iodide solution has been added to turn the color to that of strong tea. When fixing in fluids which contain sublimates there are formed so-called sublimate precipitations which not only bother the microscopist but can even lead to failures. By "iodizing", these precipitations are eliminated.

DEHYDRATION AND PARAFFIN INFILTRATION

Before you can embed the organic fragments in paraffin or celloidin, they must be completely dehydrated. To do this, substitute the water which remains in the tissues with alcohol. If you put your objects in absolute alcohol at once, however, the concentration balance between water and alcohol would cause distortions, shrinkages and even damages in the tissues. You must therefore dehydrate in stages, whereby you transfer the object from water to 30, 40, 50, 70, 80 and 95% alcohol. The lower stages (up to 70%) should last at least 4 hours, and the higher ones at least 12 hours. Here the size of the object also governs the time of immersion. By using a fixing solution which is washed out with alcohol, dehydration is already begun with the washing process. Then continue with the next higher stage.

In place of the alcohol stages you can dehydrate in one step with methyl glycol. From methyl glycol you can transfer immediately to absolute isopropyl alcohol.

As with the fixing fluid, the alcohol must come in contact with all surfaces of the organic material. Therefore use glass-wool again or, better yet, hang it in the upper layers with a bit of thread. Of course all the containers used should be sealable, or at least be covered with a piece of glass.

From the 95% alcohol transfer your objects to absolute alcohol. However, absolute ethyl alcohol is very hygroscopic (water-absorbent) so that it is difficult to keep it free of water. For this reason you will do better to use absolute isopropyl alcohol, which has other advantages as well over ethyl alcohol.

The absolute isopropyl alcohol is changed three times. Total time of immersion in absolute alcohol should be 2 to 3 days.

Because isopropyl alcohol is poorly soluble in paraffin, do not transfer the organic bits directly from absolute alcohol to paraffin, but inject an intermediary stage between alcohol and paraffin. Many experienced histologists prefer benzol as the best intermediary. The beginner will probably get better results with pure turpentine than with such quickly evaporating substances as benzol and chloroform.

Take the objects from absolute alcohol and place them in turpentine. Use the purest, guaranteed water-free turpentine. With larger, not so easily permeable objects, it is advisable to change the turpentine once or twice.

Long immersion in isopropyl alcohol and turpentine is necessary so that the pieces are well infiltrated. Unfortunately the objects also become hard and brittle after their stay in 100% alcohol and turpentine, the longer they are immersed. If you want to work correctly and value material which is supple and easily cut, insert three stages of methyl benzoate between the isopropyl alcohol and the turpentine. The objects remain in the first stage until they have sunk to the bottom (at first they float on the surface); they are then transferred to the second stage of methyl benzoate in which they remain for 12 hours (longer does no harm),

Fig. 114. Warming cabinet.

and lastly for another 12 hours in the third stage. Methyl benzoate can leave traces of water. For this purpose medium-sized objects ($\frac{1}{2}$ cm. on the edges) may be left for 24 hours in 100% isopropyl alcohol. Because methyl benzoate is also soluble in paraffin, you can shorten the immersion in turpentine: half an hour to 4 hours suffices, depending on the size of the objects. Only the most important pieces need be washed in methyl benzoate. A disadvantage of methyl benzoate is the high price.

At last you are ready for soaking in paraffin, during which the organic fragments which were soaked in turpentine should be soaked as thoroughly as possible. For this purpose you will need a paraffin stove, that is, a heated cabinet which can be held at a required temperature by an automatic thermostat. If you find that a cabinet (Fig. 114) is too expensive, you can make use of the following method:

Mount a carbon filament lamp on a retort stand so that it dips into a beaker of solid paraffin, clearing the surface by about an inch. The heat of the lamp melts the paraffin to a depth of about $\frac{1}{2}$ in. The objects to be soaked thus lie in the border zone between firm and fluid paraffin and cannot be overheated in any circumstances.

From the turpentine, transfer the material into clean paraffin heated to about 60°C., which is changed once. The immersion in fluid paraffin should be too long rather than too short, so that the object is surely soaked. Medium-sized pieces (about $\frac{1}{4}$ in. thick) may get enough in 4 to 8 hours,

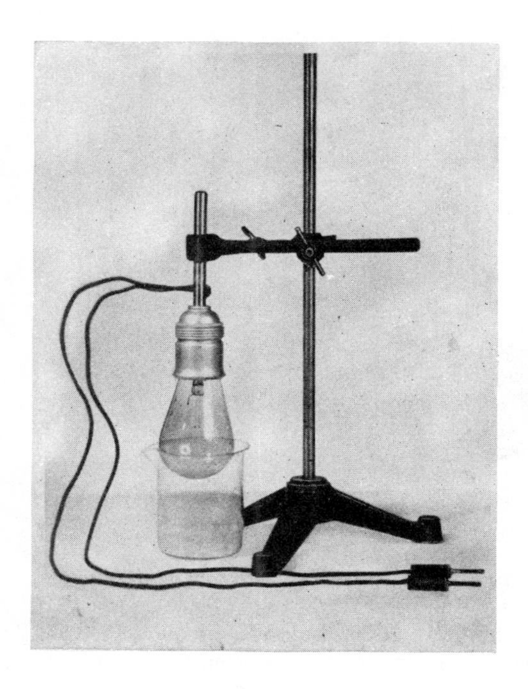

Fig. 115. Paraffin infiltration with the help of a carbon filament lamp.

but a stay of a full day in the hot paraffin does no harm, if dehydration beforehand was complete enough.

Choosing the correct kind of paraffin is not too simple. In winter use paraffin wax with a melting point of 52° to 54° C. and in summer the melting point should be 56° to 58°C.

EMBEDDING IN PARAFFIN

Once the organic fragments have become fully saturated with paraffin, you must embed them, that is, surround them with a coating of solid paraffin. For this purpose use an "embedding frame" made up of two pieces of angle metal which are held together by a clamp. Set the bedding frames on a glass pane which has been coated with glycerin and pour perfectly clean, not previously used paraffin in, drop in the objects and "position" them with a hot preparation needle to the desired height for later cutting. Then place the glass pane with the frame in a dish, in which you pour cold water until the water level is as high as the upper

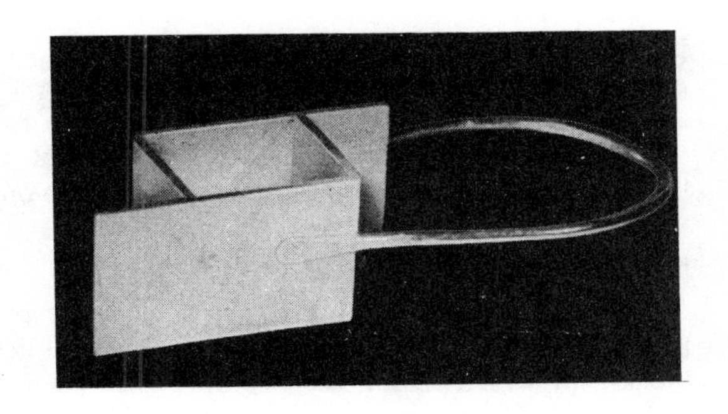

Fig. 116. Embedding frame for paraffin.

edge of the bedding frame. When the block has become completely hard, remove it and if it is not to be used at once for cutting, store in a dust-free spot. Such paraffin blocks can be saved for years without damage, while the objects would suffer some damage if preserved in 70% to 80% alcohol or formalin, and above all would lose some of their capacity for being dyed.

Instead of the embedding frame, you could also use a little glass dish which was previously coated inside with glycerin. The hard paraffin is then easily removed from the dish.

MOUNTING AND TRIMMING THE BLOCK

Before microtome cutting you must trim the paraffin block and mount it tightly on the stage. First cut off with a sharp, heated knife enough paraffin so that the object is still surrounded with a $\frac{1}{4}$ in. thickness of

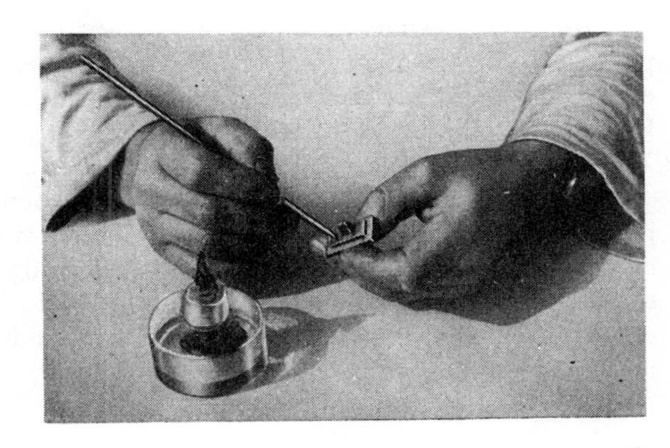

Fig. 117. Trimming the paraffin block.

paraffin. Then fasten the block to the stage with a few drops of hot paraffin (using a clamp on a bit of wood). If it rests firmly and immovably trim away again enough paraffin from the five exposed sides so that the object has only a very thin coating of paraffin around it. So that the block will not vibrate, leave a little more paraffin on the bottom. (Fig. 117.)

EMBEDDING IN CELLOIDIN

In most cases embedding in paraffin is the preferred way. Celloidin blocks are more tedious to prepare and more difficult to cut than paraffin blocks, for which reason the beginner is advised first to use the paraffin technique. There are, however, instances where celloidin is indicated, for example, for organs which have a great deal of muscle or connective tissue, for bone tissues, skin, insects, mature plant organs, all of which become brittle and are difficult to cut when embedded in paraffin.

For infiltration and embedding you will require a 2%, a 4% and an 8% solution of celloidin in a mixture of equal parts of absolute ethyl alcohol and water-free ether.

Before infiltration, transfer dehydrated organic fragments from absolute ethyl alcohol to a mixture of equal parts of absolute ethyl alcohol and water-free ether. After about 6 hours this mixture is replaced with a 2% celloidin mixture, in which the objects remain for 2 to 3 days. Then transfer them to a 4% solution for 2 to 5 days and the same time again in an 8% solution. The longer the objects are left in these celloidin stages, the better the results because celloidin permeates very slowly. Once the objects are infiltrated with celloidin, pour them into a glass dish along with the 8% solution. The dish is then placed with sulphuric acid in a desiccator. When the celloidin solution has been reduced to about half its volume, use an embedding container in a closed dish, the bottom of which has been covered with 70% alcohol. As soon as a skin has begun to form on the celloidin, which should be in a few hours, you can also pour 70% alcohol into the embedding container itself, which causes the celloidin to harden. After 1 to 2 days you can cut around and carefully lift out the objects. For additional hardening, the blocks are placed in 70% alcohol a few more days, after which they are ready for cutting.

For mounting the block, use a sliver of wood which was previously placed in an alcohol-ether solution for 10 to 15 minutes. The dried block of celloidin is fastened to this wood sliver with 8% celloidin solution and then the sliver and block are placed in 70% alcohol. In a few hours the celloidin which was used to cement the block in place has become so hard that the block can no longer be moved.

THE MICROTOME

Microtomes are made in various styles. The choice of styles is largely a matter of individual taste and the very diverse opinions of the professionals are usually dependent upon the type to which they have become accustomed.

The rocking microtome has a blade holder on a horizontal frame while the object on the object holder is usually raised by an inclined sliding plate. In some microtomes of this type, the object is raised vertically on an object holder by means of a screw column.

The rocking microtomes resemble the rotary microtomes in which the object moves past the rigid knife and after every cut a slice of predetermined thickness is pushed out.

Of course the quality of your sections is governed by the microtome blade used. Above all the blades must always be kept sharp. Because the proper honing and stropping of a microtome blade requires a considerable amount of skill, the untrained operator would do best to have his blades sharpened by a trained and equipped person. If you want to do your own work, you must do as follows:

On a fine-grained stone with a perfectly flat surface get a very thin "mud" with a rubbing-stone and water—as if you were sharpening a

Fig. 118. Rocking microtome.

razor. To the back of the blade, fasten a blade-holder[15] and then draw the blade over the stone in circular motions with the edge in front. In doing this you must be careful to hone both sides of the blade equally. Keep the holder on the blade and draw it across the strop, now of course with the edge turned away. Hones which are supple are of no use whatsoever in sharpening microtome blades. You must use a 4-sided strop mounted on a rigid base, as pictured on page 97. If you continue stropping too long, the edge will become too finely polished—crushed and compressed cuts will result. The edge should show a very fine serration when magnified 100 diameters. Sometimes it is even better to use only the coarse surface of the strop.

CUTTING WITH THE MICROTOME

All moving parts of the microtome must slide freely when cutting. You must therefore be careful that the instrument is always well oiled and that all the sticky oil layers which might adhere from previous use are wiped off.

Paraffin blocks are cut with a dry and transversely set blade. Never make a section thicker than 50 microns; otherwise the blade will be harmed and the block may be torn loose. Whether you cut rapidly, slowly or even in jerks depends on the nature of the object. Only practice and experience will be of help here.

The sections are removed from the blade with a fine camel's-hair brush. If they curl when cut, then you must try to make thinner sections or to catch each section with the brush as it is made. If this does not help, then bring the room temperature closer to the melting-point of the paraffin. If the sections are compressed, usually the blade is too flat to the cutting surface; if they break, then it is too steep. You can find the proper adjustment with the use of an adjustable blade-holder.

As a start you should amuse yourself with single sections. After having had some practice, you can start making a continuous series of sections, but this requires some experience.

If you make series sections, do not remove each section from the blade, but let them cling to the edge. Each successive section then unites with the previous one by having the edges melt together. Each successive section then pushes the ribbon for a considerable way along the blade. In this way you can get very long ribbons. (Fig. 118.)

A requisite for usable ribbons is careful embedding, a proper room temperature and thin sections.

---

[15] These may be purchased from the firms which manufacture the microtome.

Celloidin blocks are cut if possible with a highly-tilted blade and both block and blade surfaces moistened with 70% alcohol. The sections are placed in 70% alcohol. For plant specimens, especially mature plant organs, the celloidin method is sometimes better suited than the paraffin procedure.

The use of frozen sections eliminates the need for embedding the material. You can even cut them fresh; usually it is better to fix them in formalin and wash them in water first. If you do not have a freezing microtome, it is possible to place a freezing table on any ordinary microtome. The most used freezing substances are carbon dioxide or ethyl chloride. The manufacturer of your microtome will provide a carbon dioxide freezing table. For the microscope hobbyist who does not work constantly with his microtome, freezing with ethyl chloride is usually

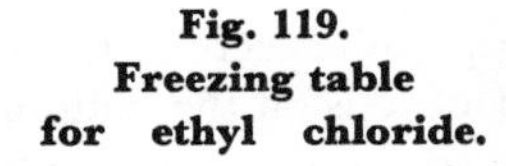
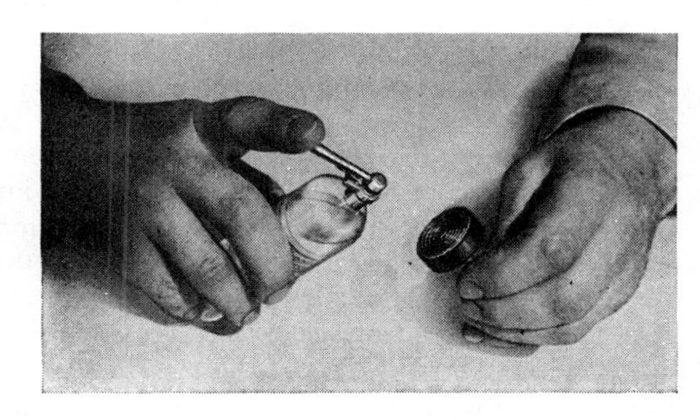

Fig. 119. Freezing table for ethyl chloride.

sufficient: provide yourself with a little metal cube which is roughened on one side and fits into the object clamp of the microtome. A moistened piece of tissue paper is placed on the roughened surface and on this the object (smaller pieces are previously embedded in a piece of raw potato). The object, after it has been embedded, is now dripped with the ethyl chloride from the wash-bottle (working the lever also permits you to drip as well as spray from this bottle). Then flood the object with ethyl chloride. Let the freezing material evaporate for a few seconds and then add an occasional drop or two—until the object is solidly frozen which can be proven by cutting. In order to avoid using too much ethyl chloride, you should choose small objects if possible. If the object is frozen fast to the stand, begin cutting, meantime always dropping a little ethyl chloride on the surface of the cut, to hinder its melting. Remember that ethyl chloride is inflammable and also has a narcotic effect. Frozen sections are placed in water and then worked as required. Because it is not possible to make the thinnest sections with the freezing method, it is primarily useful for quick diagnosis and special examinations (for instance, fat and nerve tissues).

Plant structures, especially the woody organs, often get so hard after dehydration and paraffin soaking that they cannot be cut on the microtome. In such cases the paraffin blocks are cut on one side in such a manner that the embedded tissue is laid open, and then the block is placed for several days in a mixture of glycerin and water (1:1 to 1:5). The tissues then absorb the watery solution and again become soft. Still better sometimes is a mixture of 8 parts of 80% isopropyl alcohol and 2 parts of glycerin. The infiltrated blocks must not be permitted to dry, but must always be kept in the fluid.

A hand microtome may also be used with all the methods described here. You must not—as you did previously—draw the blade through the paraffin block, but must press it in short jerks through the object.

SUBSEQUENT HANDLING OF THE SECTIONS

Paraffin sections are usually very delicate. It is best to fix them on a slide and perform all subsequent operations such as de-paraffinizing, dyeing, etc., on the slide.[16] For mounting sections, coat a clean slide thinly with Mayer's albumen[17] and line up your paraffin sections in the proper order in such a manner that the shiny underside is on the bottom. Then drop on enough distilled water so that the sections just float and then warm the slide to about 45°C. Even the best paraffin sections are sometimes wavy and must be stretched. After the warming, which is best done on the warming-bench, the sections flatten out completely. At no time during this stretching should the paraffin melt on the slide, because then the delicate tissues could tear or shrink. When all the sections have become flat, allow the water that is left to run off and place the slide in a dust-free location to dry, at best in a heating cabinet at 40°C. Only after they are completely dry can you work with the sections again.

In order that the sections stick really well, you have to get the albumen to coagulate. For this purpose, heat the slide after drying over a flame until the paraffin just begins to melt. Because the sections are now dry, melting paraffin does no more harm.

For de-paraffining, the slide is placed in a larger preparation jar filled with xylol. After 2 to 5 minutes, when the paraffin is completely absorbed, transfer for 2 to 5 minutes to absolute alcohol, then 95%, 80% and 60% alcohol, then water. If you have fixed the object in a solution containing sublimate, the sections must be iodized. For this purpose add a few drops of Lugol's solution to 80% alcohol. As iodine traces are harmful to some

[16] Single sections may also be pasted on cover-glasses.
[17] To prepare: fresh egg white is mixed well with glycerin and filtered through a wad of cotton. To prevent spoilage add sodium salicylate. Spreading this substance is best done with the fingertips.

dyes, you must remove all iodine from the sections again. For this purpose move them for a few minutes from water into a 0.25% sodium sulphate solution and then wash again with water. Because the highly diluted sodium sulphate solution cannot be stored for any length of time, it must be freshly prepared each time before use from a 10% stock solution.

Celloidin sections need not be fixed down. Transfer them from 70% alcohol to 60% and then to water.

After dyeing, the sections go through the rising alcohol stages into absolute alcohol and xylol and are then mounted in resin.

With the more complicated technique of microtome sectioning some failures cannot be avoided in the beginning.

Patience, perseverance and punctilious work are still the surest requisites for success. The satisfaction of a perfect series of sections is ample reward for all the difficulties encountered.

SAMPLE PREPARATION FOR PARAFFIN SECTIONS *

Fixing acc. to Bouin see page 144	Alcohol 70% (3 changes) 2–24 hrs. or longer	Alcohol 80% 24 hrs.	Alcohol 95% 24 hrs.
Alcohol 100% Isopropyl Alcohol 3 times 12–24 hrs.	Turpentine (1–2 changes) 12–24 hrs.	Paraffin 4–24 hrs.	Cut Paste and dry sections
Xylol 2–5 mins.	Alcohol 100% Isopropyl Alcohol 2–5 mins.	Alcohol 95% 2–5 mins.	Alcohol 80% 2–5 mins.
Alcohol 60% 2–5 mins.	Distilled Water	Hemalum 5–10 mins.	Tap Water change often 15–20 mins.
Eosin 0.1% 3–5 mins.	Distilled Water (rinse)	Alcohol 60% 2 mins.	Alcohol 80% 2 mins.
Alcohol 95% 2 mins.	Alcohol 100% Isopropyl Alcohol 5 mins.	Xylol 2–5 mins.	Resin (mounting)

* The given times are minimal times which (except for the dyeing times, the times in 100% alcohol and in turpentine) can be prolonged without damage. As a basis, an object with an edge dimension of ½ cm. was used. Larger objects require longer time for the working procedures up to sectioning.

Plant Materials

Object	Fixing and Further Handling	Dyeing	Dehydration and Mounting
Algae	Chrome acetic fluid 15 to 20 mins. Rinse in water See page 70	Alizarin viridin-chrome alum (2 hrs. to 1 day). Wash in distilled water See page 72	Place in glycerin plus water 1:6 to 1:10, allow to thicken, mount in glycerin or glycerin jelly. See page 75
Bacteria	Draw dried smears through flame, remove fats with ether See page 134	Löffler's methylene blue or carbol fuchsin. Rinse with water for 5 mins. See page 134	Allow to dry in air. Mount in resin.
Fungi	Chrome acetic fluid 15 to 30 mins. Rinse in water See page 70	Heidenhain's hematoxylin or direct deep-black See pages 91 and 103	Alcohol stages or methyl glycol; xylol; resin.
Fresh cuts of the organs of higher plants.	Chrome acetic fluid 15 to 30 mins. Rinse in water. See page 70	Leaves : Alizarin viridin-chrome alum 2 hrs or longer. Woody structures: picro-nigrosin or hematoxylin-safranin. Surface preparations : Delafield's hematoxylin or direct deep black. See pages 72 and 103	Alcohol stages, terpineol (xylol), resin. After picronigrosin dehydrate directly in 90% alcohol. Undyed cuts are mounted without dehydration in glycerin jelly.
Organs of higher plants	Formalin acetic alcohol 24 hrs. (pieces not too large, use plenty of fluid). Rinse in 60% alcohol. Razor cuttings should be bleached in Javelle water. See pages 102 and 110	See above	Alcohol stages, terpineol (xylol), resin.

Animal Materials

Object	Fixing and Further Handling	Dyeing	Dehydration and Mounting
Infusoria	Smear procedure. See page 83		Air-dried smears mounted in resin.
Water fleas, rotifers and other plankton animals, small animals and organs (whole mounts).	Formalin 1:4–1:10 (1 part 40% formalin thinned with 4–10 parts tap or spring water), hours to days. Rinse in tap water. Alcohol stages to 70% alcohol.	Grenacher's borax carmine differentiate in acetic alcohol, wash out the acid in 70% alcohol. See page 79	Alcohol stages, methyl benzoate, resin or alcohol stages and euparal.
Insects and insect organs (chitin preparations)	Fixed or not, boil in caustic potash, rinse thoroughly in acidified water, then clean water. (Better than boiling is to macerate for a time in cold caustic potash.)	—	95% alcohol, methyl benzoate, resin. See page 57
Blood	Allow smears to dry in air, fix and dye in one operation with Wright's blood stain. See page 117		Dried smears mounted in resin.
Animal organs which have been teased out	Formalin (see above) 30 mins. to days. Rinse in tap water.	Hemalum-eosin. See page 118	Alcohol stages or methyl glycol, terpineol (xylol), resin.

Index

A CATALOG OF SELECTED

DOVER BOOKS

IN ALL FIELDS OF INTEREST

A CATALOG OF SELECTED DOVER BOOKS IN ALL FIELDS OF INTEREST

THE BOOK OF BEASTS: Being a Translation from a Latin Bestiary of the Twelfth Century, T. H. White. Wonderful catalog real and fanciful beasts: manticore, griffin, phoenix, amphivius, jaculus, many more. White's witty erudite commentary on scientific, historical aspects. Fascinating glimpse of medieval mind. Illustrated. 296pp. 5⅜ × 8¼. (Available in U.S. only) 24609-4 Pa. $6.95

FRANK LLOYD WRIGHT: ARCHITECTURE AND NATURE With 160 Illustrations, Donald Hoffmann. Profusely illustrated study of influence of nature—especially prairie—on Wright's designs for Fallingwater, Robie House, Guggenheim Museum, other masterpieces. 96pp. 9¼ × 10¾. 25098-9 Pa. $7.95

FRANK LLOYD WRIGHT'S FALLINGWATER, Donald Hoffmann. Wright's famous waterfall house: planning and construction of organic idea. History of site, owners, Wright's personal involvement. Photographs of various stages of building. Preface by Edgar Kaufmann, Jr. 100 illustrations. 112pp. 9¼ × 10.
23671-4 Pa. $8.95

YEARS WITH FRANK LLOYD WRIGHT: Apprentice to Genius, Edgar Tafel. Insightful memoir by a former apprentice presents a revealing portrait of Wright the man, the inspired teacher, the greatest American architect. 372 black-and-white illustrations. Preface. Index. vi + 228pp. 8¼ × 11. 24801-1 Pa. $10.95

THE STORY OF KING ARTHUR AND HIS KNIGHTS, Howard Pyle. Enchanting version of King Arthur fable has delighted generations with imaginative narratives of exciting adventures and unforgettable illustrations by the author. 41 illustrations. xviii + 313pp. 6⅛ × 9¼. 21445-1 Pa. $6.95

THE GODS OF THE EGYPTIANS, E. A. Wallis Budge. Thorough coverage of numerous gods of ancient Egypt by foremost Egyptologist. Information on evolution of cults, rites and gods; the cult of Osiris; the Book of the Dead and its rites; the sacred animals and birds; Heaven and Hell; and more. 956pp. 6⅛ × 9¼.
22055-9, 22056-7 Pa., Two-vol. set $21.90

A THEOLOGICO-POLITICAL TREATISE, Benedict Spinoza. Also contains unfinished *Political Treatise*. Great classic on religious liberty, theory of government on common consent. R. Elwes translation. Total of 421pp. 5⅜ × 8½.
20249-6 Pa. $6.95

INCIDENTS OF TRAVEL IN CENTRAL AMERICA, CHIAPAS, AND YUCATAN, John L. Stephens. Almost single-handed discovery of Maya culture; exploration of ruined cities, monuments, temples; customs of Indians. 115 drawings. 892pp. 5⅜ × 8½. 22404-X, 22405-8 Pa., Two-vol. set $15.90

LOS CAPRICHOS, Francisco Goya. 80 plates of wild, grotesque monsters and caricatures. Prado manuscript included. 183pp. 6⅝ × 9⅝. 22384-1 Pa. $5.95

AUTOBIOGRAPHY: The Story of My Experiments with Truth, Mohandas K. Gandhi. Not hagiography, but Gandhi in his own words. Boyhood, legal studies, purification, the growth of the Satyagraha (nonviolent protest) movement. Critical, inspiring work of the man who freed India. 480pp. 5⅜ × 8½. (Available in U.S. only)
24593-4 Pa. $6.95

ILLUSTRATED DICTIONARY OF HISTORIC ARCHITECTURE, edited by Cyril M. Harris. Extraordinary compendium of clear, concise definitions for over 5,000 important architectural terms complemented by over 2,000 line drawings. Covers full spectrum of architecture from ancient ruins to 20th-century Modernism. Preface. 592pp. 7½ × 9⅜. 24444-X Pa. $15.95

THE NIGHT BEFORE CHRISTMAS, Clement Moore. Full text, and woodcuts from original 1848 book. Also critical, historical material. 19 illustrations. 40pp. 4⅝ × 6. 22797-9 Pa. $2.50

THE LESSON OF JAPANESE ARCHITECTURE: 165 Photographs, Jiro Harada. Memorable gallery of 165 photographs taken in the 1930's of exquisite Japanese homes of the well-to-do and historic buildings. 13 line diagrams. 192pp. 8⅜ × 11¼. 24778-3 Pa. $10.95

THE AUTOBIOGRAPHY OF CHARLES DARWIN AND SELECTED LETTERS, edited by Francis Darwin. The fascinating life of eccentric genius composed of an intimate memoir by Darwin (intended for his children); commentary by his son, Francis; hundreds of fragments from notebooks, journals, papers; and letters to and from Lyell, Hooker, Huxley, Wallace and Henslow. xi + 365pp. 5⅜ × 8. 20479-0 Pa. $6.95

WONDERS OF THE SKY: Observing Rainbows, Comets, Eclipses, the Stars and Other Phenomena, Fred Schaaf. Charming, easy-to-read poetic guide to all manner of celestial events visible to the naked eye. Mock suns, glories, Belt of Venus, more. Illustrated. 299pp. 5¼ × 8¼. 24402-4 Pa. $7.95

BURNHAM'S CELESTIAL HANDBOOK, Robert Burnham, Jr. Thorough guide to the stars beyond our solar system. Exhaustive treatment. Alphabetical by constellation: Andromeda to Cetus in Vol. 1; Chamaeleon to Orion in Vol. 2; and Pavo to Vulpecula in Vol. 3. Hundreds of illustrations. Index in Vol. 3. 2,000pp. 6⅛ × 9¼. 23567-X, 23568-8, 23673-0 Pa., Three-vol. set $38.85

STAR NAMES: Their Lore and Meaning, Richard Hinckley Allen. Fascinating history of names various cultures have given to constellations and literary and folkloristic uses that have been made of stars. Indexes to subjects. Arabic and Greek names. Biblical references. Bibliography. 563pp. 5⅜ × 8½. 21079-0 Pa. $8.95

THIRTY YEARS THAT SHOOK PHYSICS: The Story of Quantum Theory, George Gamow. Lucid, accessible introduction to influential theory of energy and matter. Careful explanations of Dirac's anti-particles, Bohr's model of the atom, much more. 12 plates. Numerous drawings. 240pp. 5⅜ × 8½. 24895-X Pa. $5.95

CHINESE DOMESTIC FURNITURE IN PHOTOGRAPHS AND MEASURED DRAWINGS, Gustav Ecke. A rare volume, now affordably priced for antique collectors, furniture buffs and art historians. Detailed review of styles ranging from early Shang to late Ming. Unabridged republication. 161 black-and-white drawings, photos. Total of 224pp. 8⅜ × 11¼. (Available in U.S. only) 25171-3 Pa. $13.95

VINCENT VAN GOGH: A Biography, Julius Meier-Graefe. Dynamic, penetrating study of artist's life, relationship with brother, Theo, painting techniques, travels, more. Readable, engrossing. 160pp. 5⅜ × 8½. (Available in U.S. only) 25253-1 Pa. $4.95

HOW TO WRITE, Gertrude Stein. Gertrude Stein claimed anyone could understand her unconventional writing—here are clues to help. Fascinating improvisations, language experiments, explanations illuminate Stein's craft and the art of writing. Total of 414pp. 4⅜ × 6⅜. 23144-5 Pa. $6.95

ADVENTURES AT SEA IN THE GREAT AGE OF SAIL: Five Firsthand Narratives, edited by Elliot Snow. Rare true accounts of exploration, whaling, shipwreck, fierce natives, trade, shipboard life, more. 33 illustrations. Introduction. 353pp. 5⅜ × 8½. 25177-2 Pa. $8.95

THE HERBAL OR GENERAL HISTORY OF PLANTS, John Gerard. Classic descriptions of about 2,850 plants—with over 2,700 illustrations—includes Latin and English names, physical descriptions, varieties, time and place of growth, more. 2,706 illustrations. xlv + 1,678pp. 8½ × 12¼. 23147-X Cloth. $75.00

DOROTHY AND THE WIZARD IN OZ, L. Frank Baum. Dorothy and the Wizard visit the center of the Earth, where people are vegetables, glass houses grow and Oz characters reappear. Classic sequel to *Wizard of Oz*. 256pp. 5⅜ × 8.
24714-7 Pa. $4.95

SONGS OF EXPERIENCE: Facsimile Reproduction with 26 Plates in Full Color, William Blake. This facsimile of Blake's original "Illuminated Book" reproduces 26 full-color plates from a rare 1826 edition. Includes "The Tyger," "London," "Holy Thursday," and other immortal poems. 26 color plates. Printed text of poems. 48pp. 5¼ × 7. 24636-1 Pa. $3.50

SONGS OF INNOCENCE, William Blake. The first and most popular of Blake's famous "Illuminated Books," in a facsimile edition reproducing all 31 brightly colored plates. Additional printed text of each poem. 64pp. 5¼ × 7.
22764-2 Pa. $3.50

PRECIOUS STONES, Max Bauer. Classic, thorough study of diamonds, rubies, emeralds, garnets, etc.: physical character, occurrence, properties, use, similar topics. 20 plates, 8 in color. 94 figures. 659pp. 6⅛ × 9¼.
21910-0, 21911-9 Pa., Two-vol. set $15.90

ENCYCLOPEDIA OF VICTORIAN NEEDLEWORK, S. F. A. Caulfeild and Blanche Saward. Full, precise descriptions of stitches, techniques for dozens of needlecrafts—most exhaustive reference of its kind. Over 800 figures. Total of 679pp. 8⅛ × 11. Two volumes. Vol. 1 22800-2 Pa. $11.95
Vol. 2 22801-0 Pa. $11.95

THE MARVELOUS LAND OF OZ, L. Frank Baum. Second Oz book, the Scarecrow and Tin Woodman are back with hero named Tip, Oz magic. 136 illustrations. 287pp. 5⅜ × 8½. 20692-0 Pa. $5.95

WILD FOWL DECOYS, Joel Barber. Basic book on the subject, by foremost authority and collector. Reveals history of decoy making and rigging, place in American culture, different kinds of decoys, how to make them, and how to use them. 140 plates. 156pp. 7⅞ × 10¾. 20011-6 Pa. $8.95

HISTORY OF LACE, Mrs. Bury Palliser. Definitive, profusely illustrated chronicle of lace from earliest times to late 19th century. Laces of Italy, Greece, England, France, Belgium, etc. Landmark of needlework scholarship. 266 illustrations. 672pp. 6⅛ × 9¼. 24742-2 Pa. $14.95

ILLUSTRATED GUIDE TO SHAKER FURNITURE, Robert Meader. All furniture and appurtenances, with much on unknown local styles. 235 photos. 146pp. 9 × 12. 22819-3 Pa. $8.95

WHALE SHIPS AND WHALING: A Pictorial Survey, George Francis Dow. Over 200 vintage engravings, drawings, photographs of barks, brigs, cutters, other vessels. Also harpoons, lances, whaling guns, many other artifacts. Comprehensive text by foremost authority. 207 black-and-white illustrations. 288pp. 6 × 9. 24808-9 Pa. $8.95

THE BERTRAMS, Anthony Trollope. Powerful portrayal of blind self-will and thwarted ambition includes one of Trollope's most heartrending love stories. 497pp. 5⅜ × 8½. 25119-5 Pa. $9.95

ADVENTURES WITH A HAND LENS, Richard Headstrom. Clearly written guide to observing and studying flowers and grasses, fish scales, moth and insect wings, egg cases, buds, feathers, seeds, leaf scars, moss, molds, ferns, common crystals, etc.—all with an ordinary, inexpensive magnifying glass. 209 exact line drawings aid in your discoveries. 220pp. 5⅜ × 8½. 23330-8 Pa. $4.95

RODIN ON ART AND ARTISTS, Auguste Rodin. Great sculptor's candid, wide-ranging comments on meaning of art; great artists; relation of sculpture to poetry, painting, music; philosophy of life, more. 76 superb black-and-white illustrations of Rodin's sculpture, drawings and prints. 119pp. 8⅜ × 11¼. 24487-3 Pa. $7.95

FIFTY CLASSIC FRENCH FILMS, 1912–1982: A Pictorial Record, Anthony Slide. Memorable stills from Grand Illusion, Beauty and the Beast, Hiroshima, Mon Amour, many more. Credits, plot synopses, reviews, etc. 160pp. 8¼ × 11. 25256-6 Pa. $11.95

THE PRINCIPLES OF PSYCHOLOGY, William James. Famous long course complete, unabridged. Stream of thought, time perception, memory, experimental methods; great work decades ahead of its time. 94 figures. 1,391pp. 5⅜ × 8½. 20381-6, 20382-4 Pa., Two-vol. set $23.90

BODIES IN A BOOKSHOP, R. T. Campbell. Challenging mystery of blackmail and murder with ingenious plot and superbly drawn characters. In the best tradition of British suspense fiction. 192pp. 5⅜ × 8½. 24720-1 Pa. $3.95

CALLAS: PORTRAIT OF A PRIMA DONNA, George Jellinek. Renowned commentator on the musical scene chronicles incredible career and life of the most controversial, fascinating, influential operatic personality of our time. 64 black-and-white photographs. 416pp. 5⅜ × 8¼. 25047-4 Pa. $8.95

GEOMETRY, RELATIVITY AND THE FOURTH DIMENSION, Rudolph Rucker. Exposition of fourth dimension, concepts of relativity as Flatland characters continue adventures. Popular, easily followed yet accurate, profound. 141 illustrations. 133pp. 5⅜ × 8½. 23400-2 Pa. $3.95

HOUSEHOLD STORIES BY THE BROTHERS GRIMM, with pictures by Walter Crane. 53 classic stories—Rumpelstiltskin, Rapunzel, Hansel and Gretel, the Fisherman and his Wife, Snow White, Tom Thumb, Sleeping Beauty, Cinderella, and so much more—lavishly illustrated with original 19th century drawings. 114 illustrations. x + 269pp. 5⅜ × 8½. 21080-4 Pa. $4.95

SUNDIALS, Albert Waugh. Far and away the best, most thorough coverage of ideas, mathematics concerned, types, construction, adjusting anywhere. Over 100 illustrations. 230pp. 5⅜ × 8½. 22947-5 Pa. $4.95

PICTURE HISTORY OF THE NORMANDIE: With 190 Illustrations, Frank O. Braynard. Full story of legendary French ocean liner: Art Deco interiors, design innovations, furnishings, celebrities, maiden voyage, tragic fire, much more. Extensive text. 144pp. 8⅞ × 11¾. 25257-4 Pa. $10.95

THE FIRST AMERICAN COOKBOOK: A Facsimile of "American Cookery," 1796, Amelia Simmons. Facsimile of the first American-written cookbook published in the United States contains authentic recipes for colonial favorites—pumpkin pudding, winter squash pudding, spruce beer, Indian slapjacks, and more. Introductory Essay and Glossary of colonial cooking terms. 80pp. 5⅜ × 8½. 24710-4 Pa. $3.50

101 PUZZLES IN THOUGHT AND LOGIC, C. R. Wylie, Jr. Solve murders and robberies, find out which fishermen are liars, how a blind man could possibly identify a color—purely by your own reasoning! 107pp. 5⅜ × 8½. 20367-0 Pa. $2.50

THE BOOK OF WORLD-FAMOUS MUSIC—CLASSICAL, POPULAR AND FOLK, James J. Fuld. Revised and enlarged republication of landmark work in musico-bibliography. Full information about nearly 1,000 songs and compositions including first lines of music and lyrics. New supplement. Index. 800pp. 5⅜ × 8¼. 24857-7 Pa. $15.95

ANTHROPOLOGY AND MODERN LIFE, Franz Boas. Great anthropologist's classic treatise on race and culture. Introduction by Ruth Bunzel. Only inexpensive paperback edition. 255pp. 5⅜ × 8½. 25245-0 Pa. $6.95

THE TALE OF PETER RABBIT, Beatrix Potter. The inimitable Peter's terrifying adventure in Mr. McGregor's garden, with all 27 wonderful, full-color Potter illustrations. 55pp. 4¼ × 5½. (Available in U.S. only) 22827-4 Pa. $1.75

THREE PROPHETIC SCIENCE FICTION NOVELS, H. G. Wells. *When the Sleeper Wakes, A Story of the Days to Come* and *The Time Machine* (full version). 335pp. 5⅜ × 8½. (Available in U.S. only) 20605-X Pa. $6.95

APICIUS COOKERY AND DINING IN IMPERIAL ROME, edited and translated by Joseph Dommers Vehling. Oldest known cookbook in existence offers readers a clear picture of what foods Romans ate, how they prepared them, etc. 49 illustrations. 301pp. 6⅛ × 9¼. 23563-7 Pa. $7.95

SHAKESPEARE LEXICON AND QUOTATION DICTIONARY, Alexander Schmidt. Full definitions, locations, shades of meaning of every word in plays and poems. More than 50,000 exact quotations. 1,485pp. 6½ × 9¼. 22726-X, 22727-8 Pa., Two-vol. set $29.90

THE WORLD'S GREAT SPEECHES, edited by Lewis Copeland and Lawrence W. Lamm. Vast collection of 278 speeches from Greeks to 1970. Powerful and effective models; unique look at history. 842pp. 5⅜ × 8½. 20468-5 Pa. $11.95

THE BLUE FAIRY BOOK, Andrew Lang. The first, most famous collection, with many familiar tales: Little Red Riding Hood, Aladdin and the Wonderful Lamp, Puss in Boots, Sleeping Beauty, Hansel and Gretel, Rumpelstiltskin; 37 in all. 138 illustrations. 390pp. 5⅜ × 8½. 21437-0 Pa. $6.95

THE STORY OF THE CHAMPIONS OF THE ROUND TABLE, Howard Pyle. Sir Launcelot, Sir Tristram and Sir Percival in spirited adventures of love and triumph retold in Pyle's inimitable style. 50 drawings, 31 full-page. xviii + 329pp. 6½ × 9¼. 21883-X Pa. $7.95

AUDUBON AND HIS JOURNALS, Maria Audubon. Unmatched two-volume portrait of the great artist, naturalist and author contains his journals, an excellent biography by his granddaughter, expert annotations by the noted ornithologist, Dr. Elliott Coues, and 37 superb illustrations. Total of 1,200pp. 5⅜ × 8.
Vol. I 25143-8 Pa. $8.95
Vol. II 25144-6 Pa. $8.95

GREAT DINOSAUR HUNTERS AND THEIR DISCOVERIES, Edwin H. Colbert. Fascinating, lavishly illustrated chronicle of dinosaur research, 1820's to 1960. Achievements of Cope, Marsh, Brown, Buckland, Mantell, Huxley, many others. 384pp. 5¼ × 8¼. 24701-5 Pa. $7.95

THE TASTEMAKERS, Russell Lynes. Informal, illustrated social history of American taste 1850's–1950's. First popularized categories Highbrow, Lowbrow, Middlebrow. 129 illustrations. New (1979) afterword. 384pp. 6 × 9.
23993-4 Pa. $8.95

DOUBLE CROSS PURPOSES, Ronald A. Knox. A treasure hunt in the Scottish Highlands, an old map, unidentified corpse, surprise discoveries keep reader guessing in this cleverly intricate tale of financial skullduggery. 2 black-and-white maps. 320pp. 5⅜ × 8½. (Available in U.S. only) 25032-6 Pa. $6.95

AUTHENTIC VICTORIAN DECORATION AND ORNAMENTATION IN FULL COLOR: 46 Plates from "Studies in Design," Christopher Dresser. Superb full-color lithographs reproduced from rare original portfolio of a major Victorian designer. 48pp. 9¼ × 12¼. 25083-0 Pa. $7.95

PRIMITIVE ART, Franz Boas. Remains the best text ever prepared on subject, thoroughly discussing Indian, African, Asian, Australian, and, especially, Northern American primitive art. Over 950 illustrations show ceramics, masks, totem poles, weapons, textiles, paintings, much more. 376pp. 5⅜ × 8. 20025-6 Pa. $6.95

SIDELIGHTS ON RELATIVITY, Albert Einstein. Unabridged republication of two lectures delivered by the great physicist in 1920–21. *Ether and Relativity* and *Geometry and Experience*. Elegant ideas in non-mathematical form, accessible to intelligent layman. vi + 56pp. 5⅜ × 8½. 24511-X Pa. $2.95

THE WIT AND HUMOR OF OSCAR WILDE, edited by Alvin Redman. More than 1,000 ripostes, paradoxes, wisecracks: Work is the curse of the drinking classes, I can resist everything except temptation, etc. 258pp. 5⅜ × 8½. 20602-5 Pa. $4.95

ADVENTURES WITH A MICROSCOPE, Richard Headstrom. 59 adventures with clothing fibers, protozoa, ferns and lichens, roots and leaves, much more. 142 illustrations. 232pp. 5⅜ × 8½. 23471-1 Pa. $3.95

PLANTS OF THE BIBLE, Harold N. Moldenke and Alma L. Moldenke. Standard reference to all 230 plants mentioned in Scriptures. Latin name, biblical reference, uses, modern identity, much more. Unsurpassed encyclopedic resource for scholars, botanists, nature lovers, students of Bible. Bibliography. Indexes. 123 black-and-white illustrations. 384pp. 6 × 9. 25069-5 Pa. $8.95

FAMOUS AMERICAN WOMEN: A Biographical Dictionary from Colonial Times to the Present, Robert McHenry, ed. From Pocahontas to Rosa Parks, 1,035 distinguished American women documented in separate biographical entries. Accurate, up-to-date data, numerous categories, spans 400 years. Indices. 493pp. 6½ × 9¼. 24523-3 Pa. $10.95

THE FABULOUS INTERIORS OF THE GREAT OCEAN LINERS IN HISTORIC PHOTOGRAPHS, William H. Miller, Jr. Some 200 superb photographs capture exquisite interiors of world's great "floating palaces"—1890's to 1980's: *Titanic, Ile de France, Queen Elizabeth, United States, Europa,* more. Approx. 200 black-and-white photographs. Captions. Text. Introduction. 160pp. 8⅞ × 11¾. 24756-2 Pa. $9.95

THE GREAT LUXURY LINERS, 1927–1954: A Photographic Record, William H. Miller, Jr. Nostalgic tribute to heyday of ocean liners. 186 photos of Ile de France, Normandie, Leviathan, Queen Elizabeth, United States, many others. Interior and exterior views. Introduction. Captions. 160pp. 9 × 12. 24056-8 Pa. $10.95

A NATURAL HISTORY OF THE DUCKS, John Charles Phillips. Great landmark of ornithology offers complete detailed coverage of nearly 200 species and subspecies of ducks: gadwall, sheldrake, merganser, pintail, many more. 74 full-color plates, 102 black-and-white. Bibliography. Total of 1,920pp. 8⅝ × 11¼. 25141-1, 25142-X Cloth. Two-vol. set $100.00

THE SEAWEED HANDBOOK: An Illustrated Guide to Seaweeds from North Carolina to Canada, Thomas F. Lee. Concise reference covers 78 species. Scientific and common names, habitat, distribution, more. Finding keys for easy identification. 224pp. 5⅜ × 8½. 25215-9 Pa. $6.95

THE TEN BOOKS OF ARCHITECTURE: The 1755 Leoni Edition, Leon Battista Alberti. Rare classic helped introduce the glories of ancient architecture to the Renaissance. 68 black-and-white plates. 336pp. 8⅜ × 11¼. 25239-6 Pa. $14.95

MISS MACKENZIE, Anthony Trollope. Minor masterpieces by Victorian master unmasks many truths about life in 19th-century England. First inexpensive edition in years. 392pp. 5⅜ × 8½. 25201-9 Pa. $8.95

THE RIME OF THE ANCIENT MARINER, Gustave Doré, Samuel Taylor Coleridge. Dramatic engravings considered by many to be his greatest work. The terrifying space of the open sea, the storms and whirlpools of an unknown ocean, the ice of Antarctica, more—all rendered in a powerful, chilling manner. Full text. 38 plates. 77pp. 9¼ × 12. 22305-1 Pa. $4.95

THE EXPEDITIONS OF ZEBULON MONTGOMERY PIKE, Zebulon Montgomery Pike. Fascinating first-hand accounts (1805–6) of exploration of Mississippi River, Indian wars, capture by Spanish dragoons, much more. 1,088pp. 5⅜ × 8½. 25254-X, 25255-8 Pa. Two-vol. set $25.90

DEGAS: An Intimate Portrait, Ambroise Vollard. Charming, anecdotal memoir by famous art dealer of one of the greatest 19th-century French painters. 14 black-and-white illustrations. Introduction by Harold L. Van Doren. 96pp. 5⅜ × 8½.
25131-4 Pa. $4.95

PERSONAL NARRATIVE OF A PILGRIMAGE TO ALMANDINAH AND MECCAH, Richard Burton. Great travel classic by remarkably colorful personality. Burton, disguised as a Moroccan, visited sacred shrines of Islam, narrowly escaping death. 47 illustrations. 959pp. 5⅜ × 8½. 21217-3, 21218-1 Pa., Two-vol. set $19.90

PHRASE AND WORD ORIGINS, A. H. Holt. Entertaining, reliable, modern study of more than 1,200 colorful words, phrases, origins and histories. Much unexpected information. 254pp. 5⅜ × 8½. 20758-7 Pa. $5.95

THE RED THUMB MARK, R. Austin Freeman. In this first Dr. Thorndyke case, the great scientific detective draws fascinating conclusions from the nature of a single fingerprint. Exciting story, authentic science. 320pp. 5⅜ × 8½. (Available in U.S. only) 25210-8 Pa. $6.95

AN EGYPTIAN HIEROGLYPHIC DICTIONARY, E. A. Wallis Budge. Monumental work containing about 25,000 words or terms that occur in texts ranging from 3000 B.C. to 600 A.D. Each entry consists of a transliteration of the word, the word in hieroglyphs, and the meaning in English. 1,314pp. 6⅞ × 10.
23615-3, 23616-1 Pa., Two-vol. set $31.90

THE COMPLEAT STRATEGYST: Being a Primer on the Theory of Games of Strategy, J. D. Williams. Highly entertaining classic describes, with many illustrated examples, how to select best strategies in conflict situations. Prefaces. Appendices. xvi + 268pp. 5⅜ × 8½. 25101-2 Pa. $5.95

THE ROAD TO OZ, L. Frank Baum. Dorothy meets the Shaggy Man, little Button-Bright and the Rainbow's beautiful daughter in this delightful trip to the magical Land of Oz. 272pp. 5⅜ × 8. 25208-6 Pa. $5.95

POINT AND LINE TO PLANE, Wassily Kandinsky. Seminal exposition of role of point, line, other elements in non-objective painting. Essential to understanding 20th-century art. 127 illustrations. 192pp. 6½ × 9¼. 23808-3 Pa. $4.95

LADY ANNA, Anthony Trollope. Moving chronicle of Countess Lovel's bitter struggle to win for herself and daughter Anna their rightful rank and fortune—perhaps at cost of sanity itself. 384pp. 5⅜ × 8½. 24669-8 Pa. $8.95

EGYPTIAN MAGIC, E. A. Wallis Budge. Sums up all that is known about magic in Ancient Egypt: the role of magic in controlling the gods, powerful amulets that warded off evil spirits, scarabs of immortality, use of wax images, formulas and spells, the secret name, much more. 253pp. 5⅜ × 8½. 22681-6 Pa. $4.50

THE DANCE OF SIVA, Ananda Coomaraswamy. Preeminent authority unfolds the vast metaphysic of India: the revelation of her art, conception of the universe, social organization, etc. 27 reproductions of art masterpieces. 192pp. 5⅜ × 8½.
24817-8 Pa. $5.95

CHRISTMAS CUSTOMS AND TRADITIONS, Clement A. Miles. Origin, evolution, significance of religious, secular practices. Caroling, gifts, yule logs, much more. Full, scholarly yet fascinating; non-sectarian. 400pp. 5⅜ × 8½.
23354-5 Pa. $6.95

THE HUMAN FIGURE IN MOTION, Eadweard Muybridge. More than 4,500 stopped-action photos, in action series, showing undraped men, women, children jumping, lying down, throwing, sitting, wrestling, carrying, etc. 390pp. 7⅞ × 10⅝.
20204-6 Cloth. $21.95

THE MAN WHO WAS THURSDAY, Gilbert Keith Chesterton. Witty, fast-paced novel about a club of anarchists in turn-of-the-century London. Brilliant social, religious, philosophical speculations. 128pp. 5⅜ × 8½. 25121-7 Pa. $3.95

A CEZANNE SKETCHBOOK: Figures, Portraits, Landscapes and Still Lifes, Paul Cezanne. Great artist experiments with tonal effects, light, mass, other qualities in over 100 drawings. A revealing view of developing master painter, precursor of Cubism. 102 black-and-white illustrations. 144pp. 8¾ × 6⅜. 24790-2 Pa. $5.95

AN ENCYCLOPEDIA OF BATTLES: Accounts of Over 1,560 Battles from 1479 B.C. to the Present, David Eggenberger. Presents essential details of every major battle in recorded history, from the first battle of Megiddo in 1479 B.C. to Grenada in 1984. List of Battle Maps. New Appendix covering the years 1967–1984. Index. 99 illustrations. 544pp. 6½ × 9¼. 24913-1 Pa. $14.95

AN ETYMOLOGICAL DICTIONARY OF MODERN ENGLISH, Ernest Weekley. Richest, fullest work, by foremost British lexicographer. Detailed word histories. Inexhaustible. Total of 856pp. 6½ × 9¼.
21873-2, 21874-0 Pa., Two-vol. set $17.00

WEBSTER'S AMERICAN MILITARY BIOGRAPHIES, edited by Robert McHenry. Over 1,000 figures who shaped 3 centuries of American military history. Detailed biographies of Nathan Hale, Douglas MacArthur, Mary Hallaren, others. Chronologies of engagements, more. Introduction. Addenda. 1,033 entries in alphabetical order. xi + 548pp. 6½ × 9¼. (Available in U.S. only)
24758-9 Pa. $13.95

LIFE IN ANCIENT EGYPT, Adolf Erman. Detailed older account, with much not in more recent books: domestic life, religion, magic, medicine, commerce, and whatever else needed for complete picture. Many illustrations. 597pp. 5⅜ × 8½.
22632-8 Pa. $8.95

HISTORIC COSTUME IN PICTURES, Braun & Schneider. Over 1,450 costumed figures shown, covering a wide variety of peoples: kings, emperors, nobles, priests, servants, soldiers, scholars, townsfolk, peasants, merchants, courtiers, cavaliers, and more. 256pp. 8⅜ × 11¼. 23150-X Pa. $9.95

THE NOTEBOOKS OF LEONARDO DA VINCI, edited by J. P. Richter. Extracts from manuscripts reveal great genius; on painting, sculpture, anatomy, sciences, geography, etc. Both Italian and English. 186 ms. pages reproduced, plus 500 additional drawings, including studies for *Last Supper, Sforza* monument, etc. 860pp. 7⅞ × 10¾. (Available in U.S. only) 22572-0, 22573-9 Pa., Two-vol. set $31.90

THE ART NOUVEAU STYLE BOOK OF ALPHONSE MUCHA: All 72 Plates from "Documents Decoratifs" in Original Color, Alphonse Mucha. Rare copyright-free design portfolio by high priest of Art Nouveau. Jewelry, wallpaper, stained glass, furniture, figure studies, plant and animal motifs, etc. Only complete one-volume edition. 80pp. 9⅝ × 12¼. 24044-4 Pa. $9.95

ANIMALS: 1,419 COPYRIGHT-FREE ILLUSTRATIONS OF MAMMALS, BIRDS, FISH, INSECTS, ETC., edited by Jim Harter. Clear wood engravings present, in extremely lifelike poses, over 1,000 species of animals. One of the most extensive pictorial sourcebooks of its kind. Captions. Index. 284pp. 9 × 12. 23766-4 Pa. $9.95

OBELISTS FLY HIGH, C. Daly King. Masterpiece of American detective fiction, long out of print, involves murder on a 1935 transcontinental flight—"a very thrilling story"—NY Times. Unabridged and unaltered republication of the edition published by William Collins Sons & Co. Ltd., London, 1935. 288pp. 5⅜ × 8½. (Available in U.S. only) 25036-9 Pa. $5.95

VICTORIAN AND EDWARDIAN FASHION: A Photographic Survey, Alison Gernsheim. First fashion history completely illustrated by contemporary photographs. Full text plus 235 photos, 1840–1914, in which many celebrities appear. 240pp. 6½ × 9¼. 24205-6 Pa. $6.95

THE ART OF THE FRENCH ILLUSTRATED BOOK, 1700–1914, Gordon N. Ray. Over 630 superb book illustrations by Fragonard, Delacroix, Daumier, Doré, Grandville, Manet, Mucha, Steinlen, Toulouse-Lautrec and many others. Preface. Introduction. 633 halftones. Indices of artists, authors & titles, binders and provenances. Appendices. Bibliography. 608pp. 8⅜ × 11¼. 25086-5 Pa. $24.95

THE WONDERFUL WIZARD OF OZ, L. Frank Baum. Facsimile in full color of America's finest children's classic. 143 illustrations by W. W. Denslow. 267pp. 5⅜ × 8½. 20691-2 Pa. $7.95

FRONTIERS OF MODERN PHYSICS: New Perspectives on Cosmology, Relativity, Black Holes and Extraterrestrial Intelligence, Tony Rothman, et al. For the intelligent layman. Subjects include: cosmological models of the universe; black holes; the neutrino; the search for extraterrestrial intelligence. Introduction. 46 black-and-white illustrations. 192pp. 5⅜ × 8½. 24587-X Pa. $7.95

THE FRIENDLY STARS, Martha Evans Martin & Donald Howard Menzel. Classic text marshalls the stars together in an engaging, non-technical survey, presenting them as sources of beauty in night sky. 23 illustrations. Foreword. 2 star charts. Index. 147pp. 5⅜ × 8½. 21099-5 Pa. $3.95

FADS AND FALLACIES IN THE NAME OF SCIENCE, Martin Gardner. Fair, witty appraisal of cranks, quacks, and quackeries of science and pseudoscience: hollow earth, Velikovsky, orgone energy, Dianetics, flying saucers, Bridey Murphy, food and medical fads, etc. Revised, expanded In the Name of Science. "A very able and even-tempered presentation."—The New Yorker. 363pp. 5⅜ × 8. 20394-8 Pa. $6.95

ANCIENT EGYPT: ITS CULTURE AND HISTORY, J. E Manchip White. From pre-dynastics through Ptolemies: society, history, political structure, religion, daily life, literature, cultural heritage. 48 plates. 217pp. 5⅜ × 8½. 22548-8 Pa. $5.95

SIR HARRY HOTSPUR OF HUMBLETHWAITE, Anthony Trollope. Incisive, unconventional psychological study of a conflict between a wealthy baronet, his idealistic daughter, and their scapegrace cousin. The 1870 novel in its first inexpensive edition in years. 250pp. 5⅜ × 8½. 24953-0 Pa. $5.95

LASERS AND HOLOGRAPHY, Winston E. Kock. Sound introduction to burgeoning field, expanded (1981) for second edition. Wave patterns, coherence, lasers, diffraction, zone plates, properties of holograms, recent advances. 84 illustrations. 160pp. 5⅜ × 8¼. (Except in United Kingdom) 24041-X Pa. $3.95

INTRODUCTION TO ARTIFICIAL INTELLIGENCE: SECOND, ENLARGED EDITION, Philip C. Jackson, Jr. Comprehensive survey of artificial intelligence—the study of how machines (computers) can be made to act intelligently. Includes introductory and advanced material. Extensive notes updating the main text. 132 black-and-white illustrations. 512pp. 5⅜ × 8½. 24864-X Pa. $8.95

HISTORY OF INDIAN AND INDONESIAN ART, Ananda K. Coomaraswamy. Over 400 illustrations illuminate classic study of Indian art from earliest Harappa finds to early 20th century. Provides philosophical, religious and social insights. 304pp. 6⅜ × 9⅜. 25005-9 Pa. $9.95

THE GOLEM, Gustav Meyrink. Most famous supernatural novel in modern European literature, set in Ghetto of Old Prague around 1890. Compelling story of mystical experiences, strange transformations, profound terror. 13 black-and-white illustrations. 224pp. 5⅜ × 8½. (Available in U.S. only) 25025-3 Pa. $6.95

ARMADALE, Wilkie Collins. Third great mystery novel by the author of *The Woman in White* and *The Moonstone*. Original magazine version with 40 illustrations. 597pp. 5⅜ × 8½. 23429-0 Pa. $9.95

PICTORIAL ENCYCLOPEDIA OF HISTORIC ARCHITECTURAL PLANS, DETAILS AND ELEMENTS: With 1,880 Line Drawings of Arches, Domes, Doorways, Facades, Gables, Windows, etc., John Theodore Haneman. Sourcebook of inspiration for architects, designers, others. Bibliography. Captions. 141pp. 9 × 12. 24605-1 Pa. $7.95

BENCHLEY LOST AND FOUND, Robert Benchley. Finest humor from early 30's, about pet peeves, child psychologists, post office and others. Mostly unavailable elsewhere. 73 illustrations by Peter Arno and others. 183pp. 5⅜ × 8½. 22410-4 Pa. $4.95

ERTÉ GRAPHICS, Erté. Collection of striking color graphics: *Seasons, Alphabet, Numerals, Aces* and *Precious Stones*. 50 plates, including 4 on covers. 48pp. 9⅜ × 12¼. 23580-7 Pa. $6.95

THE JOURNAL OF HENRY D. THOREAU, edited by Bradford Torrey, F. H. Allen. Complete reprinting of 14 volumes, 1837–61, over two million words; the sourcebooks for *Walden*, etc. Definitive. All original sketches, plus 75 photographs. 1,804pp. 8½ × 12¼. 20312-3, 20313-1 Cloth., Two-vol. set $120.00

CASTLES: THEIR CONSTRUCTION AND HISTORY, Sidney Toy. Traces castle development from ancient roots. Nearly 200 photographs and drawings illustrate moats, keeps, baileys, many other features. Caernarvon, Dover Castles, Hadrian's Wall, Tower of London, dozens more. 256pp. 5⅜ × 8¼. 24898-4 Pa. $6.95